SUPERFICIAL FUNGAL INFECTIONS

SUPERFICIAL FUNGAL INFECTIONS

Other titles in the *New Clinical Applications* Series:

Dermatology (Series Editor Dr J. Verbov)
Dermatological Surgery
Talking Points No. 1
Treatment No. 1

Cardiology (Series Editor Dr D. Longmore)
Cardiology Screening

Gastroenterology (Series Editor Prof. S. Matern)

Rheumatology (Series Editors Dr J. J. Calabro and Dr W. Carson Dick)

NEW
CLINICAL
APPLICATIONS
DERMATOLOGY

SUPERFICIAL FUNGAL INFECTIONS

Editor

JULIAN L. VERBOV

JP, MD, FRCP, FIBiol

Consultant Dermatologist
Royal Liverpool Hospital,
Liverpool, UK

MTP PRESS LIMITED
a member of the KLUWER ACADEMIC PUBLISHERS GROUP
LANCASTER / BOSTON / THE HAGUE / DORDRECHT

Published in the UK and Europe by
MTP Press Limited
Falcon House
Lancaster, England

British Library Cataloguing in Publication Data

Superficial fungal infections.— (New
clinical applications. Dermatology)
1. Dermatomycoses
I. Verbov, Julian II. Series
616.9'69 RL765

ISBN 0-85200-943-7
ISBN 0-85200-823-6 Series

Published in the USA by
MTP Press
A division of Kluwer Boston Inc
190 Old Derby Street
Hingham, MA 02043, USA

Typeset and printed by Butler & Tanner Ltd,
Frome and London

CONTENTS

LIST OF AUTHORS

Y. M. Clayton, PhD
Senior Lecturer,
Dept. of Medical Mycology,
The Institute of Dermatology,
5 Lisle St.,
London
WC2H 7BJ

R. J. Hay, DM, MRCP, MRCPath
Consultant Dermatologist,
Medical Mycology and Tropical Dermatology Unit,
Department of Medical Microbiology,
London School of Hygiene and Tropical Medicine,
Keppel St.,
London
WC1E 7HT

D. T. Roberts, MB, ChB, FRCP(Glas.)
Consultant Dermatologist,
Western Infirmary,
Royal Hospital for Sick Children,
and Southern General Hospital,
Glasgow

S. O. B. Roberts, MA, MB, FRCP
Consultant Dermatologist,
Addenbrooke's Hospital,
Cambridge
CB2 2QQ

F. A. Ive, MB, FRCP
Consultant Dermatologist,
Dryburn Hospital,
Durham
DH1 5TW

C. M. E. Rowland Payne, MB, MRCP
Senior Registrar in Dermatology,
Westminster Hospital,
Dean Ryle St.,
London
SW1P 2AP

SERIES EDITOR'S FOREWORD

Superficial fungal infections of the skin are a common everyday problem in both hospital and family practice. In this volume, experts in the field contribute an authoritative discussion of this important topic and the book concludes with a stimulating update on the mystery of seborrhoeic dermatitis.

JULIAN VERBOV

Royal Liverpool Hospital,
Prescot Street,
Liverpool L7 8XP

ABOUT THE EDITOR

Dr Julian Verbov is Consultant Dermatologist to Liverpool Health Authority and Clinical Lecturer in Dermatology at the University of Liverpool.

He is a member of the British Association of Dermatologists representing the British Society for Paediatric Dermatology on its Executive Committee. He is a Council Member of the Dermatology Section of the Royal Society of Medicine, and a Committee Member of the North of England Dermatological Society.

He is a Fellow of the Zoological Society of London and a Member of the Society of Authors. He is a popular speaker and author of some 180 publications. His special interests include paediatric dermatology, inherited disorders, dermatoglyphics, pruritus ani, therapeutics, and medical humour. He organizes the British Postgraduate Course in Paediatric Dermatology and is a Member of the Editorial Boards of both *Clinical and Experimental Dermatology and Pediatric Dermatology*.

1

SCALP RINGWORM (tinea capitis)

YVONNE M. CLAYTON

Scalp ringworm (tinea capitis) is primarily a disease of children, although it may rarely be seen in adults. Since the Second World War scalp ringworm has become uncommon in Europe and the Western Hemisphere, but it is still very widespread in developing countries. During this time, too, there have been marked changes in respect to which fungus species is the predominant cause of infection; for example, the virtual disappearance of *Microsporum audouinii* from Europe and the emergence of *M. canis* in its place.

CAUSATIVE FUNGI

Dermatophytes may be divided into three groups – anthropophilic, zoophilic or geophilic – according to whether the natural habitat is on humans, animals or in the soil. Fungi in all three groups are able to invade hair (Table 1.1). *Trichophyton rubrum* is an uncommon cause

TABLE 1.1 Dermatophytes causing scalp ringworm

Anthropophilic species	*Zoophilic species*	*Geophilic species*
Microsporum audouinii	*M. canis*	*M. gypseum*
M. ferrugineum		
Trichophyton tonsurans	*T. mentagrophytes*	
T. violaceum	*T. verrucosum*	
T. soudanense		
T. schoenleinii		

of scalp ringworm and two other anthropophilic species, *Epidermophyton floccosum* and *T. concentricum*, are unable to parasitize hairs.

TYPE OF HAIR INVASION

Two main forms of scalp ringworm may be differentiated on mycological grounds according to the type of hair invasion that occurs:

Ectothrix infection

The spores here are predominantly outside the hair shaft and are either small, 2–3 microns, and arranged in masses (small-spored ectothrix) or large, 3–5 or 5–10 microns, when they occur usually in straight chains on the surface of the hair.

Small-spored ectothrix invasion of hair is produced by the *Microsporum* species (Fig. 1.1) and the larger ectothrix by *T. mentagrophytes* and *T. verrucosum*.

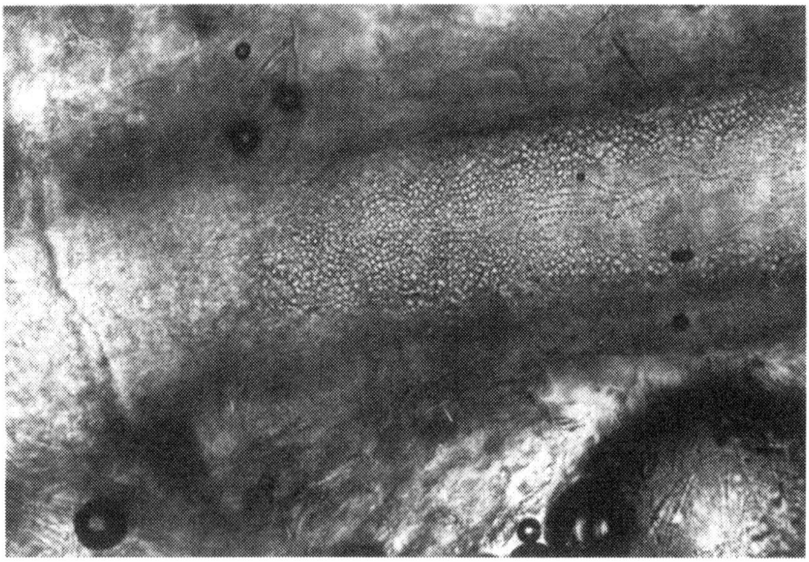

FIGURE 1.1 Small-spored ectothrix invasion of hair by *M. canis* (× 162)

Endothrix infection

Spores are confined to the hair shaft and this type of invasion is produced by *T. tonsurans*, *T. violaceum* and *T. soudanense* (Fig. 1.2).

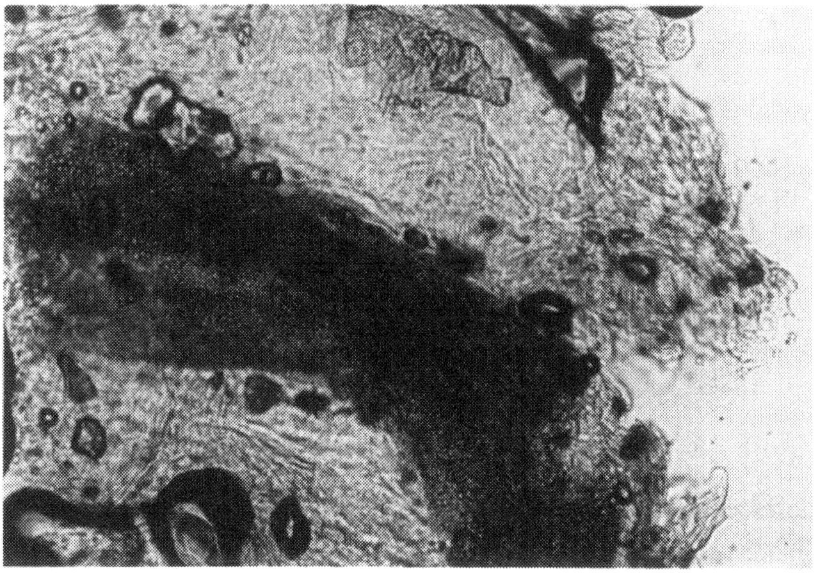

FIGURE 1.2 Endothrix invasion of hair by *T. violaceum* (×162)

Favus, which is a chronic form of tinea capitis caused by *T. schoenleinii*, shows a characteristic type of endothrix invasion in which hyphae and air spaces are found in the hair but very few spores are present (Fig. 1.3).

FLUORESCENCE

A bright green fluorescence of infected hairs seen under an ultraviolet (Wood's) lamp is a feature of hair invaded by *Microsporum* species. Apart from favus hairs which fluoresce a dull green, *Trichophyton*-infected hairs show no fluorescence. The responsible fluorescent substance is formed only when the keratin of the hair is invaded and its presence in detectable amounts under a Wood's lamp may depend on the degree to which the hairs are infected. The chemical nature of the

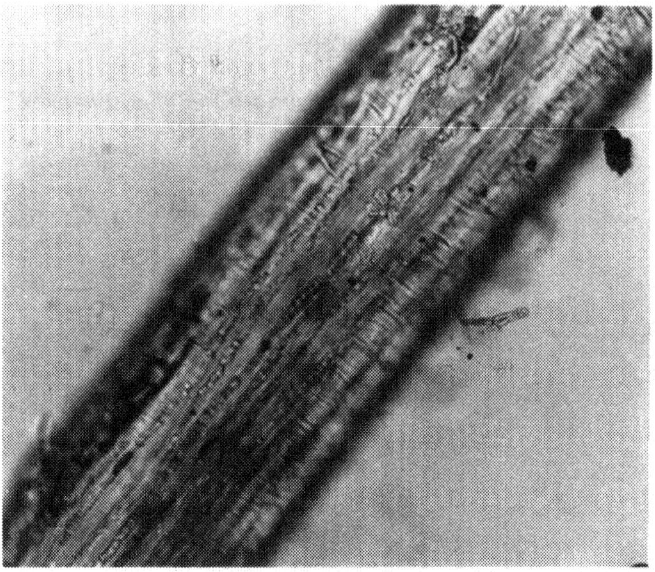

FIGURE 1.3 Favus hair showing air spaces and hyphae (×162)

fluorescent material produced has been studied[1] but is not yet fully defined.

PATHOGENESIS

The pathogenesis of experimentally induced and naturally acquired scalp ringworm was studied by Kligman[2,3]. The initial infection of scalp skin is followed by growth of the hyphae down into the potential space between the hair shaft and the follicle walls. The hair is then invaded at about mid-follicle level. The intrapilary hyphae grow down towards the hair bulb but stop at the upper limit of the keratogenous zone, known as Adamson's fringe. Hyphae within the hair are carried upwards by hair growth but the newly keratinized shaft is invaded by fungus as soon as it is being formed at the rate of about 0.3 mm daily.

Ectothrix infection

When the invading fungus is one that produces this type of infection, the intrapilary hyphae emerge through the hair surface and fragment into masses of spores (Fig. 1.4). The infected hair grows upwards and usually breaks off a few millimetres above the scalp surface.

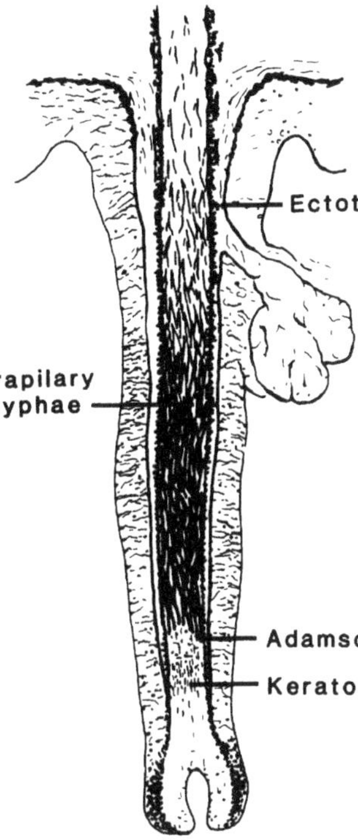

FIGURE 1.4 Diagram illustrating fungal invasion of hair (after Kligman[3])

Endothrix infection

The intrapilary hyphae break up to form spores which become rounded and fill the cortex of the hair. Because of the greater degree of shaft damage, hairs break off close to the scalp surface.

In favus, the hyphae grow irregularly down the hair shaft and do not fragment into masses of spores, with the result that the shaft is not weakened and long lengths of hair are carried beyond the follicle by the growth of the hair. Degeneration of the hyphae in the infected hair leads to the formation of elongated air spaces.

CLINICAL APPEARANCES OF SCALP RINGWORM

These are to a great extent dependent on the fungus responsible for the infection, but there is always hair loss with varying degrees of scaling and erythema.

Ectothrix infections

When *M. audouinii* is the causative fungus, slowly growing circular patches with partial hair loss occur in the scalp which show only mild erythema and scaling (Fig. 1.5). More inflammatory lesions with a

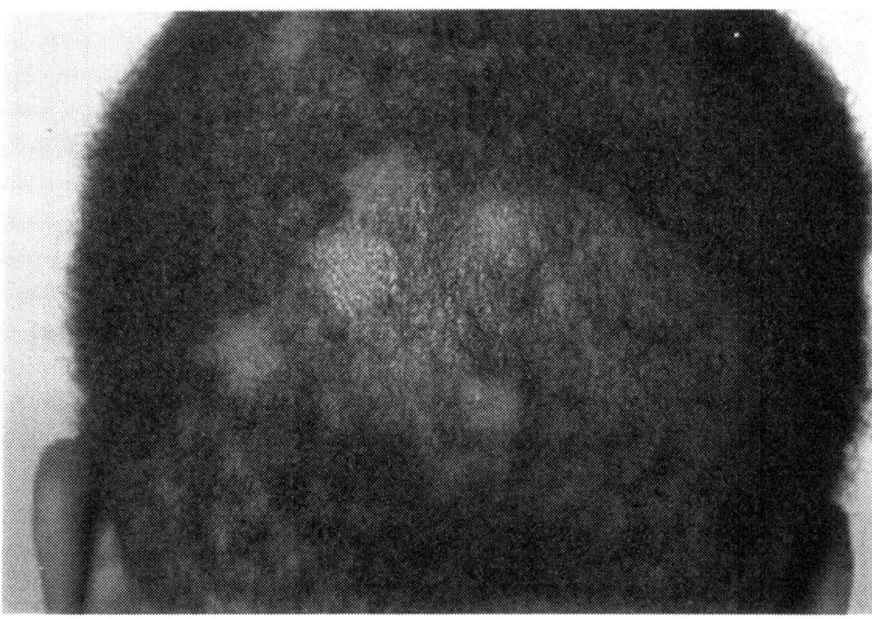

FIGURE 1.5 Scalp ringworm caused by *M. audouinii*

greater degree of scaling usually occur when *M. canis* is responsible (Fig. 1.6). However, these differences are relative and infections produced by these two fungi are often clinically indistinguishable; both show bright green infected hairs under a Wood's lamp. To determine which fungus is the cause of infection, hairs must be cultured.

Ectothrix infections caused by the zoophilic species *T. mentagrophytes*

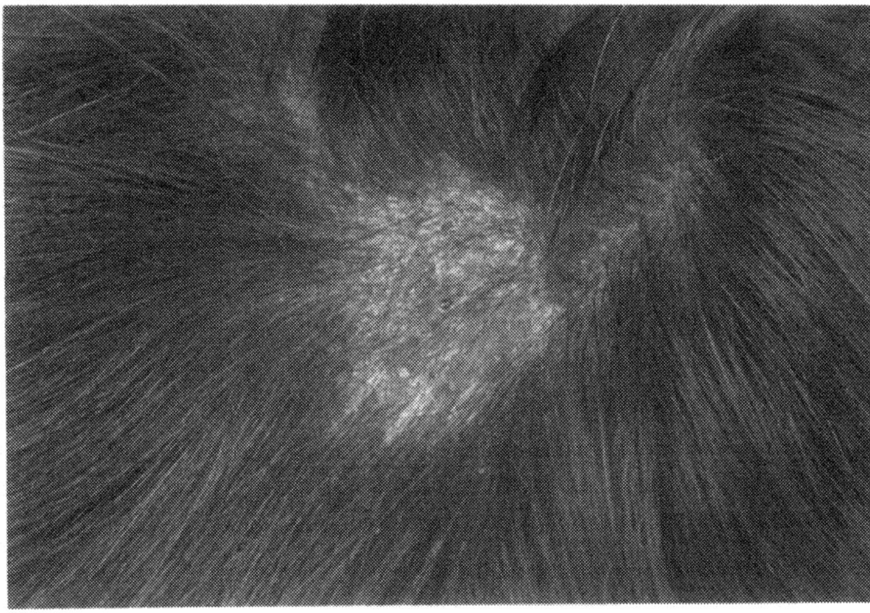

FIGURE 1.6 Scalp ringworm caused by *M. canis*

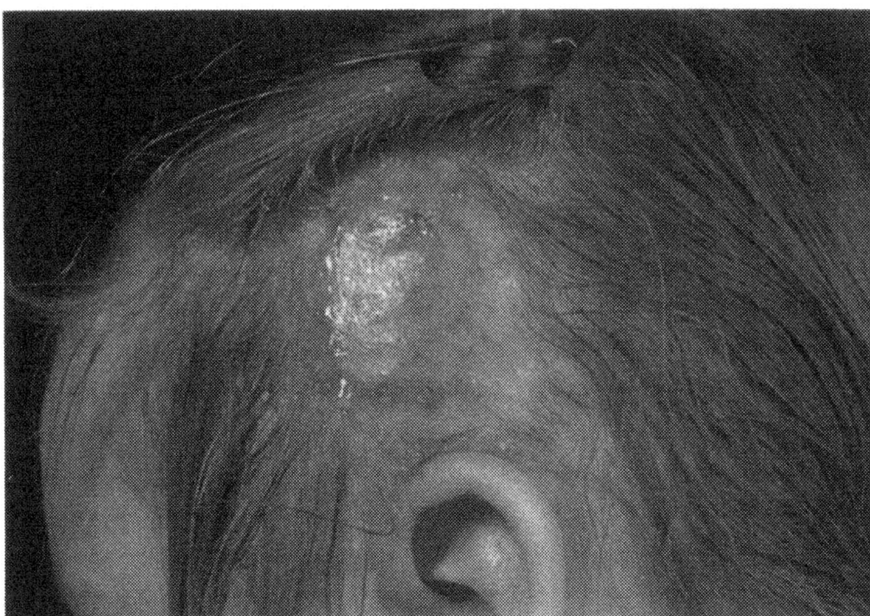

FIGURE 1.7 Kerion caused by *T. mentagrophytes*

and *T. verrucosum* are often more inflammatory than those produced by *M. canis*. Single lesions are typical with swollen and pustular areas containing loose hairs, referred to as a kerion (Fig. 1.7). No fluorescence is seen under a Wood's lamp.

Endothrix infections

Circular patches of alopecia develop with some scaling and erythema. The infected hairs break off at the scalp surface and in dark-haired subjects give a 'black-dot' appearance (Fig. 1.8). In some cases widespread intense scaling with generalized erythema occurs which

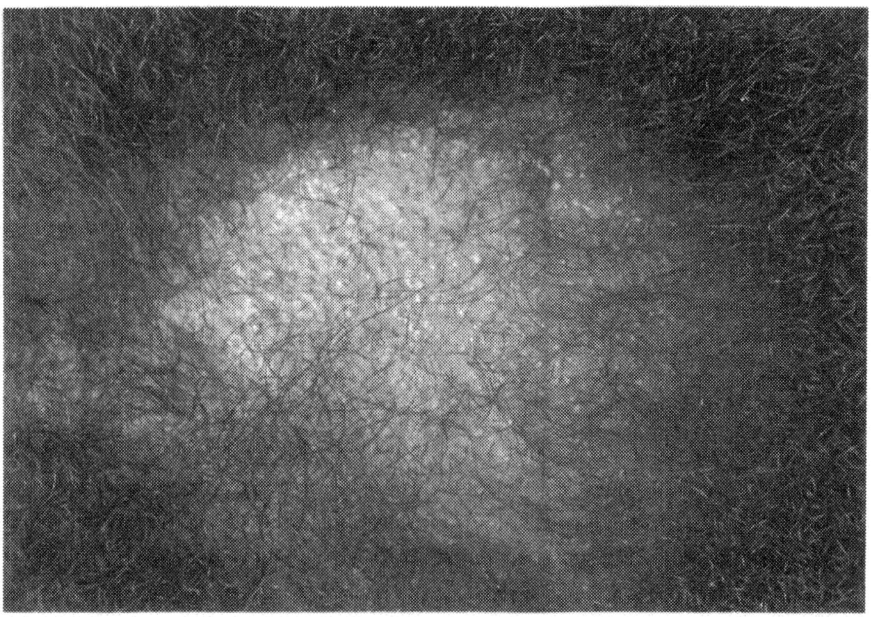

FIGURE 1.8 Endothrix caused by *T. tonsurans*

resembles seborrhoeic dermatitis; this is followed by progressive hair loss. A kerion may also develop. There is no fluorescence under a Wood's lamp.

Favus is characterized by yellowish cup-shaped crusted lesions (known as scutula) which develop around follicular openings (Fig. 1.9). The scutula, which may have a mousy odour, consist of

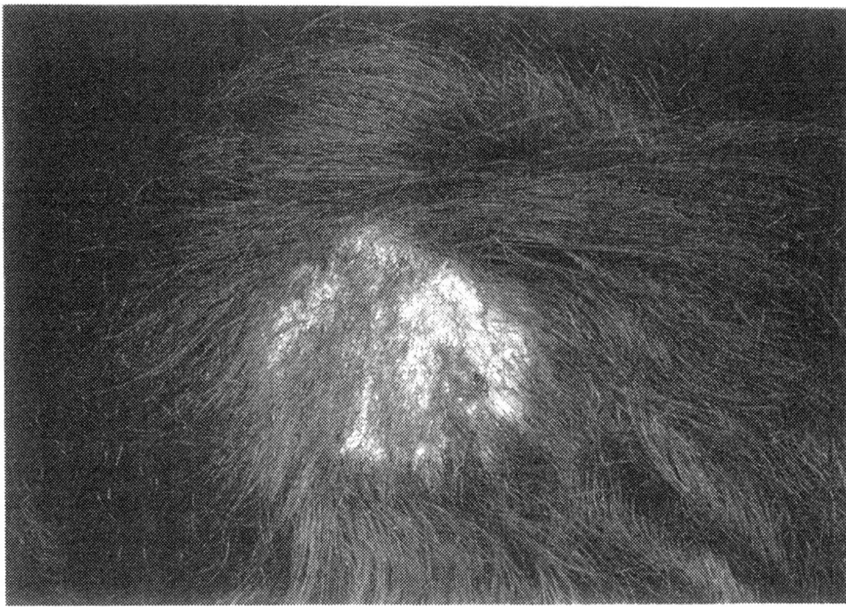

FIGURE 1.9 Favus

hyphae and spores. The infective process frequently produces considerable scarring and alopecia. Unlike other forms of scalp ringworm, infection may persist into adult life. Infected hairs show a dull green fluorescence under Wood's lamp.

HOST RESPONSE

Serum antibody

Circulating antibodies have been demonstrated in patients with natural infections. Complement-fixing antibodies were detected in 78% (18 of 23) patients with infections caused by *M. audouinii*[4]. Patients with favus have been shown to have circulating antibodies[5]. Grappel *et al.*[6], using charcoal agglutination tests, studied 130 sera from patients with tinea capitis caused by either *M. audouinii*, *T. mentagrophytes* or *T. tonsurans* and found that 40–60% gave positive reactions depending on the species. Svejgaard[7], using counterimmunoelectrophoresis, found specific antidermatophyte antibodies present in

67–99% of tinea capitis cases, predominantly in those complicated by kerion.

There is little evidence that serum antibody which has been detected indicates anything more than infection by the fungus. In particular there is little to suggest that these antibodies confer immunity to the disease. However, the report that antibody is more common where kerion is present may indicate that this is correlated with host response to the disease, although what part it plays is unclear.

Cell-mediated reactions

Data based on delayed hypersensitivity reactions to trichophytin antigens in patients with dermatophytoses suggest that development of delayed hypersensitivity is correlated with immunity to infection, and that the lack of this type of reaction is associated with susceptibility to chronic infection[8]. With respect to tinea capitis, another report[9] indicates that positive delayed hypersensitivity skin tests are strongly correlated with a kerion type of infection. It was found that 15 of 16 children with kerion caused by *T. tonsurans* gave a delayed (48 h) reaction to trichophytin with >10 mm induration, while none of the 36 children with non-inflammatory tinea capitis had a similar response. The authors suggest that the kerion is a manifestation of the cell-mediated immune response.

As is the case with circulating antibodies, there is little data on the role of cell-mediated reactions in the host response to tinea capitis.

SCALP RINGWORM IN ADULTS

Scalp ringworm is unusual after puberty. At puberty there is generally spontaneous resolution of infection, even in untreated cases, suggesting that some protective factors then become operative. This resolution could be due to changes in the amount of sebum excreted after puberty[10] or to changes in the composition of sebum with an increase in fungistatic acids[11]. However, other, perhaps immunological, factors may play a role.

Favus does, however, occur in adults. The initial infection probably starts in childhood in most cases[12] but unlike other types of scalp ringworm does not tend to clear spontaneously at puberty. Cases frequently occur in families with different generations being infected[13,14].

Adult scalp ringworm caused by other dermatophytes is uncommon. A report of tinea capitis in the adult and adolescent by Pipkin[15] in the United States, and covering a period of 20 years, described 39 cases in adults and 18 in adolescents. In London only 25 cases in adults were diagnosed over a 16-year period[16]. In the first report, the fungi most frequently implicated were *T. tonsurans* (42 cases) and *M. canis* (8). Of the 25 cases seen in London, *T. tonsurans* was isolated from 9, *M. canis* from 8, *M. audouinii* from 4, and *T. violaceum* and *T. soudanense* each from 2 cases.

One of the striking features of scalp ringworm in adults is the preponderance of females infected; in the two studies quoted, 37 of 39 and 23 of 25 patients were adult females. The reason for this difference is unknown.

EPIDEMIOLOGY

Periodic changing patterns in the prevalence of a particular dermatophyte as the cause of scalp ringworm have occurred throughout the world. Before 1950, *M. audouinii* was the dominant cause of tinea capitis in Britain, although in London and other cities, infections by *M. audouinii* and *M. canis* were approximately equal in number[17]. Since 1970 the number of *M. audouinii* infections has declined steadily and very few cases are now recorded in Britain[18]. *Microsporum canis* has become the commonest cause of fluorescent scalp ringworm in Britain and in other European countries. In Japan and Korea, where prior to the 1970s *M. ferrugineum* predominated, *M. canis* is now isolated more frequently. *Microsporum canis* is also the species most commonly isolated from cases of tinea capitis in many South American countries[19] and also in Australia and New Zealand[20,21].

In the United States, *M. audouinii* caused widespread outbreaks during the 1940s[22,23] but since that time has steadily declined, and in a survey covering the years 1979–81 accounted for only 0.3% of the total isolations of dermatophytes[24]. One reason put forward for this decline was the exceptional success of griseofulvin in treating scalp infections caused by *M. audouinii*.

Apart from one or two areas, for example in Arizona where *M. canis* predominates[25], *T. tonsurans* is now the commonest causative agent of tinea capitis in the United States. For example, it was found that in Chicago this species accounted for 96% of all cases seen from 1978 to 1980[26]. *Trichophyton tonsurans* is endemic in Mexico, Puerto Rico and other Latin American countries and it was thought that immigration from these countries probably led to the spread of infection in the United States. The majority of cases of *T. tonsurans* scalp ringworm in the United States occur in the black population. The infection is now rarely seen in Caucasians but the reason for this is unknown.

In other parts of the world, other species are prevalent. For example in India[27] and Kenya[28] and parts of North Africa, *T. violaceum* is the commonest cause of tinea capitis. *Trichophyton soudanense* and two other anthropophilic species, *T. yaoundei* and *T. gourvilii*, are agents of tinea capitis in Central and West Africa.

Favus occurs mostly in parts of the Middle East and North Africa but is occasionally encountered in other countries[14,29].

The variety of fungi isolated from 107 cases of scalp ringworm diagnosed at St John's Hospital for Diseases of the Skin, London, during 1984 is shown in Table 1.2. As expected, *M. canis* was the predominant species accounting for 63% of all isolation. *Trichophyton violaceum*, the next most frequently isolated species (21%) was mainly grown from Indian children and *T. soudanense* from West African children. Although this generation of immigrants were born in Britain,

TABLE 1.2 Fungi isolated from scalp ringworm in London in 1984

Species	*No. of cases*
Microsporum canis	67
M. audouinii	2
Trichophyton violaceum	23
T. soudanense	7
T. tonsurans	1
T. verrucosum	7
Total	107

they still tend to be infected by fungi endemic in their home countries. These species are regularly imported into Britain but the indigenous population rarely becomes infected. *Microsporum audouinii* (2 cases) and *T. tonsurans* (1) are now seen only very sporadically in Britain.

Since 1960, a variety of *M. canis* which produces glabrous colonies on primary culture has been isolated from cases of scalp and body ringworm seen in the London area[30]. This glabrous form now represents one-third of the scalp isolates of *M. canis*. It has also been reported from other parts of England and Wales and isolated from infected cats and dogs.

Sex distribution

In the past, tinea capitis has always been recorded as being more common in boys than girls[18,22,31]; as many as three boys being infected to every girl. It has been suggested that boys were more frequently affected because shorter haircuts allow easier access for infecting spores. In recent years, whichever fungus species causes the scalp infection, cases are more equally distributed among the sexes[20]. Of 111 cases of tinea capitis seen in London in 1973, 76 occurred in boys compared to 35 in girls, whereas in 1984, 50 cases were seen in boys compared to 57 in girls. In one survey of *T. tonsurans* infections[26] girls outnumbered boys by about 3 : 2. Longer hairstyles for boys, which possibly provide a natural barrier to infection, may have influenced these changes.

TRANSMISSION AND SOURCE OF INFECTION

The spread of tinea capitis caused by anthropophilic species such as *M. audouinii* and *T. tonsurans* may be either from direct person-to-person contact or indirect through fomites such as combs, hairbrushes and clothing. Contact in schools is possibly the most important factor in the spread of scalp ringworm.

During outbreaks of tinea capitis in either day or boarding schools it is not difficult to isolate the infecting fungus from sources other than the infected children. Mackenzie[32] was able to grow *T. tonsurans* from brushes, combs, bedding and dormitory floors in a residential school where 21 pupils had either tinea capitis or tinea corporis. In an outbreak caused by *M. audouinii* in a primary school[33], it was possible to grow the fungus from dust samples taken from the cloakrooms and assembly hall and from the floors, blackboard ledges and desks in the classrooms.

Tinea capitis caused by a zoophilic dermatophyte such as *M. canis* usually develops as a result of direct contact with the infected cat or dog, and the majority of cases tend to occur in family units harbouring the animal. Outbreaks in schools caused by zoophilic species are uncommon but it should be realized that an infected child is a potential source of infection for others. During a survey of 24 000 schoolchildren attending primary day schools in the Greater London Area, brush samples from scalps[34] were taken to assess the carriage rates of fungi among clinically normal children[35]. At the time the samples were taken, any cases of scalp ringworm which had previously been diagnosed at the schools were excluded. During the investigation further cases of infection were often diagnosed and in these classes up to 31% of the children yielded the same fungus species from their scalp samples (Table 1.3). This scalp carriage is believed to represent contamination of spores shed from an infected child. It is usually transient, for once clinical cases are removed from the school the carrier rate decreases and disappears. Nevertheless, the findings illustrate that there is widespread dissemination of fungal spores from children who have scalp ringworm and some of the asymptomatic carriers may later develop

TABLE 1.3 Scalp carriage of fungi on schoolchildren[33]

Dermatophyte species	Percentage of carriers found by brush cultures	
	In classes with clinical cases	*In classes with no cases*
M. audouinii	12.1	0.8
M. ferrugineum	31.0	1 0
M. canis	12.7	0.5
T. tonsurans	17.8	1.4
T. soudanense	8.6	0.4

clinical infections. Ive[36] investigated a school in Nigeria for the incidence of tinea capitis and found that on repeated sampling asymptomatic carriers continued to be positive 4 months after the removal of diagnosed cases, and by this time several had developed clinical lesions.

MYCOLOGICAL DIAGNOSIS

Microscopy

Hairs may be examined by placing them in a drop of 30% potassium hydroxide on a slide and covering with a coverslip. No heat should be applied since it is important not to destroy the arrangement of the spores (Figs 1.1 to 1.3).

Culture

Sabouraud's dextrose agar containing a broad-spectrum antibacterial antibiotic (e.g. chloramphenicol) and cycloheximide (to exclude contaminating fungi) is the medium most commonly used. Incubation at room temperature is adequate but 26–28 °C is preferable. Cultures can usually be identified after 10–14 days incubation.

Scalps may be sampled for culture by using a brush sampling technique[34]. A commercially available massage brush (Fig. 1.10) is pressed into a Petri dish containing medium after vigorously brushing the scalp. This is a useful procedure when large numbers of children have to be examined. Children with scalp ringworm grow fungal colonies from nearly every point of the brush (Fig. 1.11). The isolation of a few colonies (1 to 10) from children with no signs of infection but who have been in contact with cases of scalp ringworm indicates a carrier state.

TREATMENT

Oral griseofulvin is still the treatment of choice for scalp ringworm. The recommended dosage for children is 10 mg/kg body weight a day,

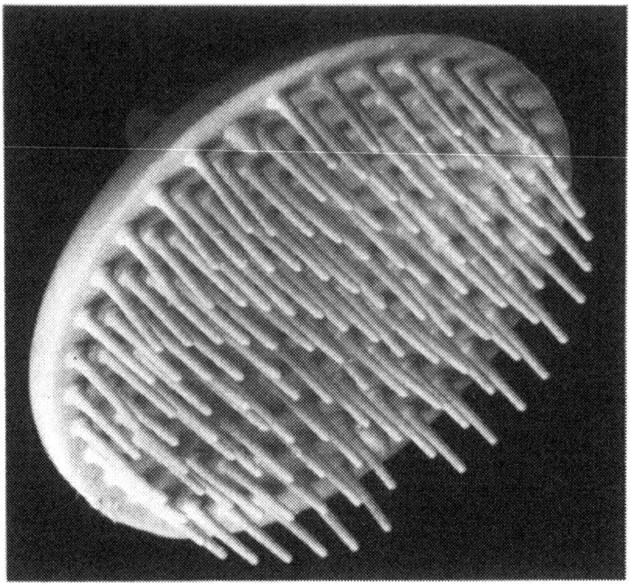

FIGURE 1.10 Brush used for sampling scalps

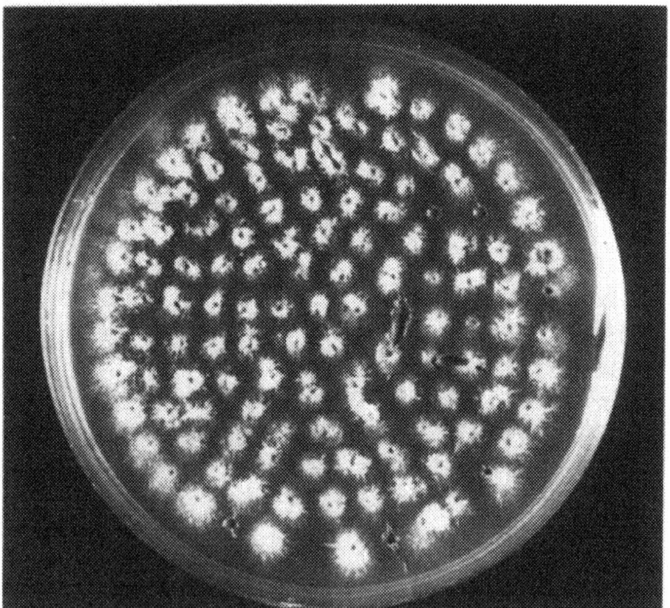

FIGURE 1.11 Culture of *T. mentagrophytes* grown from brush sample from child with scalp ringworm

given either as a single dose or divided into two, and is best taken after food. Six to 8 weeks therapy is usually effective but treatment should be continued until there is good regrowth of hair and Wood's lamp examination is negative in cases of fluorescent scalp ringworm.

A single dose of griseofulvin has been used successfully to treat cases of scalp ringworm. Friedman *et al.*[37] treated 22 children infected by *M. audouinii* with a single oral dose of 3 g and obtained cure in 21, while in a study carried out in the Congo[38] cure was obtained in 93% of 72 children after a single dose of 1.5 g. This form of therapy has not gained general acceptance, but it may have a place in developing countries where daily administration of the drug may be expensive and difficult to monitor.

ACKNOWLEDGEMENTS

Figures 1.5 to 1.9 are the copyright of the Institute of Dermatology.

REFERENCES

1. Chattaway, F. W. and Barlow, A. J. E. (1954). The fluorescent materials produced *in vivo* by certain dermatophytes. *J. Gen. Microbiol.*, **11,** 506–511
2. Kligman, A. M. (1952). The pathogenesis of tinea capitis due to *Microsporum audouinii* and *M. canis*. *J. Invest. Dermatol.*, **18,** 231–246
3. Kligman, A. M. (1955). Tinea capitis due to *M. audouinii* and *M. canis*. *Arch. Dermatol.*, **71,** 313–337
4. Kolmer, J. A. and Strickler, A. (1915). Complement fixation in parasitic skin diseases. *J. Am. Med. Assoc.*, **64,** 800–804
5. Grappel, S. F., Blank, F. and Bishop, C. T. (1971). Circulating antibodies in human favus. *Dermatologica*, **143,** 271–276
6. Grappel, S. F., Blank, F. and Bishop, C. T. (1972). Circulating antibodies in dermatophytosis. *Dermatologica*, **144**, 1–11
7. Svejgaard, E. (1985). Immunologic investigations of dermatophytes and dermatophytosis. *Sem. Dermatol.*, **4,** 201–221
8. Jones, H. E., Reinhardt, J. H. and Rinaldi, M. G. (1973). A clinical, mycological, and immunological survey for dermatophytosis. *Arch Dermatol.*, **108,** 61–65
9. Rasmussen, J. E. and Ahmed, A. R. (1978). Trichophytin reactions in children with tinea capitis. *Arch. Dermatol.*, **114,** 371–372
10. Kligman, A. M. and Ginsberg, D. (1950). Immunity of the adult scalp to infection with *Microsporum audouinii*. *J. Invest. Dermatol.*, **14,** 345–358
11. Rothman, S., Smiljanic, A., Shapiro, A. L. and Weitkamp, A. W. (1947). The spontaneous cure of tinea capitis in puberty. *J. Invest. Dermatol.*, **8,** 81–98

12. Roberts, S. O. B. and Mackenzie, D. W. R. (1979). *Textbook of Dermatology*, 3rd Edn. Vol. 1, pp. 796–797. (Oxford: Blackwell Scientific Publications)
13. Hansell, J. and Partridge, B. M. (1955). Favus: a report of seven related cases. *Br. Med. J.*, **1**, 1510–1511
14. Gibson, J. R. and Pegum, J. S. (1983). Favus – a report of two cases. *Clin. Exp. Dermatol.*, **8**, 421–423
15. Pipkin, J. L. (1952). Tinea capitis in the adult and adolescent. *Arch. Dermatol.*, **66**, 9–40
16. Ridley, C. M. (1979). Tinea capitis in an elderly woman. *Clin. Exp. Dermatol.*, **4**, 247–249
17. Walker, J. (1950). The dermatophytoses of Great Britain. Report of a three year survey. *Br. J. Dermatol.*, **62**, 239–251
18. Clayton, Y. M. (1977). Epidemiological aspects of dermatophyte infections. *Curr. Ther. Res. Clin. Exp.*, **22**, 2–9
19. Philpot, C. M. (1978). Geographical distribution of dermatophytes; a review. *J. Hyg. Camb.*, **80**, 301–313
20. McAleer, R. (1980). Fungal infections of the scalp in Western Australia. *Sabouraudia*, **18**, 185–190
21. Allred, B. J. (1982). Dermatophyte prevalence in Wellington, New Zealand. *Sabouraudia*, **20**, 75–77
22. Benedek, T. and Felsher, I. M. (1944). Epidemiology of tinea capitis. *Arch. Dermatol.*, **49**, 120–123
23. Franks, A. G. and Taschdjian, C. L. (1956). Incidence and etiology of tinea capitis. *Arch. Dermatol.*, **74**, 349–351
24. Sinski, J. T. and Flouras, K. (1984). A survey of dermatophytes isolated from human patients in the United States from 1979 to 1981 with chronological listings of worldwide incidence of five dermatophytes often isolated in the United States. *Mycopathologia*, **85**, 97–120
25. Sinski, J. T. (1982). Dermatophytosis in Tucson, Arizona, 1967–1981. *Ariz. Med. J.*, **39**, 638–640
26. Bronson, D. M., Desai, D. R., Barsky, S. and McMillen Foley, S. (1983). An epidemic of infection with *Trichophyton tonsurans* revealed in a 20-year survey of fungal infections in Chicago. *J. Am. Acad. Dermatol.*, **8**, 322–330
27. Kamalam, A. and Thambiah, A. S. (1980). Tinea capitis as an endemic disease in Madras. *Mycopathologia*, **71**, 45–51
28. Verhagen, A. R. H. B., Maniar, S. H. and Vanbreuseghem, R. (1969). Dermatophytosis in Kenya. *Trans. R. Soc. Trop. Med. Hyg.*, **3**, 275–282
29. Jacyk, W. K., Baran, E., Lawande, R. V. and Walow, B. (1982). Tinea capitis in Northern Nigeria. *Mykosen*, **25**, 221–226
30. Midgley, G. (1981). A glabrous *Microsporum canis* in Greater London. *Sabouraudia*, **19**, 71–77
31. Blank, F., Mann, S. J. and Reale, R. A. (1974). Distribution of dermatophytoses according to age, ethnic group and sex. *Sabouraudia*, **12**, 352–361
32. Mackenzie, D. W. R. (1961). The extra-human occurrence of *Trichophyton tonsurans* in a residential school. *Sabouraudia*, **1**, 58–64
33. Alexander, S., Clayton, Y. M. and Noble, W. C. (1965). Tinea capitis in a primary school. *Br. J. Dermatol.*, **77**, 373–376
34. Midgley, G. and Clayton, Y. M. (1972). Distribution of dermatophytes and Candida spores in the environment. *Br. J. Dermatol.*, **86**, 69–77

35. Clayton, Y. M. and Midgley, G. (1971). The increasing variety of ringworm fungi currently affecting London schoolchildren. *C.R. des Communications V Congress ISHAM Paris*, pp. 98–99
36. Ive, F. A. (1966). The carrier stage of tinea capitis in Nigeria. *Br. J. Dermatol.*, **78**, 219–221
37. Friedman, L., Derbes, V. J. and Tromovitch, T. A. (1960). Single dose therapy of tinea capitis. *Arch. Dermatol.*, **82**, 415–418
38. Vanbreuseghem, R., Gatti, F. and Ceballos, J. A. (1970). Mass treatment of scalp ringworm by a single dose of griseofulvin. *Int. J. Dermatol.*, **9**, 59–63

2

CHRONIC DERMATOPHYTE INFECTIONS

R. J. HAY

INTRODUCTION

The dermatophytes are mould fungi which are adapted to living in the stratum corneum or specialized keratinized structures derived from epidermis such as hair or nails[1]. They can invade human tissue from a number of different sources: soil, animals or other infected humans. The mode of transmission of dermatophytes and their subsequent penetration into stratum corneum have been the subject of surprisingly little attention over the years. However, a number of factors are thought to be important in determining a successful transfer, including the viability and virulence of organisms, site of invasion, host genotype, composition of sebum and skin surface carbon dioxide tension. Once infection has been established the subsequent course of the disease varies considerably from a short-lived process to an indolent and recalcitrant infection lasting for months or years. In the more indolent infections subsequent recovery may be spontaneous or achieved via specific chemotherapy, but often relapse or persistence occurs despite the use of antifungal drugs at doses normally considered to be adequate. Many dermatophyte infections may last for over a year, and often considerably longer if left untreated. In practice it is seldom possible to find significant numbers of patients who have not received some form of treatment, but chronic infections often remain refractory to therapy. They include infections of various skin sites, particularly the feet, hair or nails, and most are caused by anthropophilic fungi, those originating from another human.

HOST RESISTANCE

Normally resistance to infection is determined by two main factors: innate and acquired immunity[2–4]. Innate forms of resistance include the presence of serum factors which are inhibitory to the growth of dermatophytes such as unsaturated transferrin. Medium chain length fatty acids in sebum are similarly inhibitory and their presence may account for the virtual restriction of scalp ringworm infections to pre-pubertal children. A further defence mechanism involves increased epidermal cell turnover, which leads to enhanced shedding of the stratum corneum including invading fungi. The mechanism underlying this process is not clear but immune mechanisms appear to amplify the increased cell replication in experimental candida infections. Few studies have examined the role of phagocytic mechanisms in defence against dermatophytes. For instance, although their precise role has not been investigated fully, neutrophils show chemotaxis towards dermatophyte antigens[5], a process which depends on activation of the alternative pathway of complement, and it is likely that polymorphs play a role in defence against follicular invasion by fungi, where they may account for the major part of the cell infiltrates seen in histological sections.

The role of acquired immunity in defence against dermatophytes has been explored in some depth. Fungal antigens diffuse through the epidermis and are probably processed up by Langerhans cells[6]. The development of specific antibodies to dermatophytes can be demonstrated in many patients[7], although their prevalence and titre varies considerably with the methods used and the extent and duration of infection. The greatest responses are seen in certain chronic infections such as favus. However, attempts to transfer resistance to reinfection with antibody containing serum in experimental infected mice have not been successful[8]. Conversely there is strong evidence that the appearance of specifically activated T-lymphocytes is critical in the recovery process. The development of cell-mediated immunity in experimentally infected guinea pigs or humans[9], or humans with naturally acquired dermatophytosis, correlates with the start of healing[10]. In the model system using experimentally infected mice, referred to above, transfer of T-lymphocytes from previously infected donor animals into sublethally irradiated mice conveys resistance to infection on the recipients[8]. Removal of T-lymphocytes with anti-Thy-I monoclonal antiserum abrogates this response. It is still not clear by what mechanism the development of cell-mediated immunity aids the recovery from infection, but it appears to be a critical event.

As the development of immunity during a dermatophyte infection appears to be a usual event, chronic infections are likely to reflect some failure in the normal processes of resistance.

CHRONIC DERMATOPHYTOSIS

It is difficult to define a chronic infection in precise terms because of the considerable variations in the course of infection and the effects of chemotherapy. However, here chronic dermatophyte infections are taken to be those infections whose untreated course is over a year, or which fail to respond to oral antifungal therapy given for at least 6 months in adequate doses (griseofulvin 10–15 mg kg^{-1} daily, ketoconazole at least 200 mg daily), or which relapse at least twice within 3 months of stopping therapy. Given this definition the major chronic infections are onychomycosis, anthropophilic scalp ringworm, dry type *Trichophyton rubrum* infections and certain forms of tinea corporis, particularly those caused by *T. rubrum* or *T. concentricum*.

ONYCHOMYCOSIS

Nail infections with dermatophytes represent a common source of chronic disease[11,12]. With the exception of superficial white onychomycosis, most dermatophyte nail infections do not show spontaneous improvement. Even after 2 years of griseofulvin therapy only about 40% of patients with toe nail infections in a large series were uninfected at follow-up 1 year later. Those with finger nail infections fared better, but 22% remained infected. The usual explanation for this poor response is that the normal mediators of host resistance cannot be mobilized in nail keratin and adequate treatment is entirely dependent on the concentration of drug throughout the nail remaining above its inhibitory concentration for the organism concerned. Factors such as poor drug binding contribute to low nail levels, particularly at the distal margin. A further factor which encourages chronicity is the development of irregularities of the nail bed underlying the nail which lead to the formation of pockets or canals on the undersurface of the nail which can be filled with necrotic keratin and large amounts of fungus[12]. To the observer these appear as white or yellow longitudinal streaks in the nail plate (Fig. 2.1). As they may develop as the nail

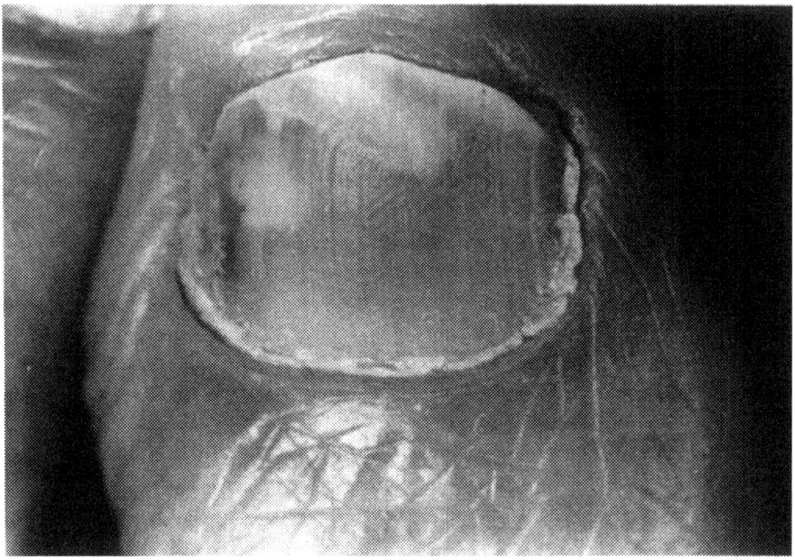

FIGURE 2.1 Onychomycosis due to *T. interdigitale* – note the streaks running in the longitudinal plane which indicate residual lacunae of infection

grows out during therapy it is thought that the drug penetration into these lacunae is poor.

Effective treatment for onychomycosis has to address several problems. These include ensuring adequate penetration of drug into the relevant site and maintenance of adequate levels there, prevention of the development of antifungal resistance or tolerance, and an approach to the eradication of sequestered foci of infection within or beneath the nail.

ANTHROPOPHILIC SCALP RINGWORM

The second group of potentially chronic dermatophyte infections are the anthropophilic scalp infections. The main organisms responsible are discussed elsewhere in this volume (see Chapter 1). While anthropophilic *Microsporum* infections such as *M. audouinii* may present problems, most attention has focused on *Trichophyton* infections such as those caused by *T. tonsurans*, *T. violaceum* and *T. soudanense*. Of these *T. tonsurans* has been best studied[13]. It has been shown, for instance, that this organism may cause indolent non-inflammatory patches of hair loss or highly inflamed lesions on the scalp. Children with the latter pattern are much more likely to show positive delayed-type hypersensitivity to intradermal trichophytin, indicating the development of cell-mediated immunity, and to show subsequent healing. In addition to overt infections it is possible for human scalp ringworm fungi to be carried in clinically normal scalps. With *T. soudanense*, for instance, children who are carriers may subsequently develop infection, although some lose the organism at follow-up re-examination without ever developing the disease. The evidence from *T. tonsurans* suggests that the long course of this infection seen in some patients may result from failure to develop adequate immunity, and that when this appears recovery is swift. Human favus provides a contrast. Here the infecting organism, *T. schoenleinii*, elicits an antibody reaction and the inflammatory response, as seen by development of crusts containing significant numbers of neutrophils, plus follicular destruction, is well developed. However, little is known of cell-mediated immunity in this infection. It is possible that this is defective and that the inflammatory response, although intense, is not capable of destroying the fungus.

While anthropophilic tinea capitis may last for considerable periods there is less problem in treating individual cases, although some patients with *T. tonsurans* may respond slowly to griseofulvin. In this disease, therefore, the major therapeutic challenge is the detection of cases and ensuring eradication by mass treatment. In many endemic areas the social, economic and medical implications of scalp ringworm are small compared to other diseases and of necessity prevention takes a low priority in health care programmes.

'DRY-TYPE' *Trichophyton rubrum* INFECTION

In the experience of most dermatologists practising in industrialized societies the management of 'dry-type' *T. rubrum* infections presents a major challenge. While many patients also have onychomycosis the infection is often refractory to treatment even if the nail involvement is ignored. In a survey of 106 patients with griseofulvin-unresponsive dermatophyte infections seen in the UK, 94% were caused by *T. rubrum*[14]. The most common sites of involvement in decreasing order of frequency were soles, toe webs, toe nails, palms, finger nails. In a control group of responsive dermatophyte infections the main sites of involvement were toe webs, groins and trunk. Of these infections 62% were caused by *T. rubrum*. In the treatment-unresponsive group infections caused by other dermatophytes, mainly *T. interdigitale* or *M. canis*, were associated with hereditary palmoplantar keratoderma (tylosis) in the former instance, and chronic mucocutaneous candidosis in the latter. The main predisposing factor in those with *T. rubrum* seen in 45% of chronic cases was a personal or family history of atopy, mainly hay fever or asthma. Other predisposing abnormalities included ichthyosis or keratoderma (7 patients), steroid therapy (3), systemic lupus erythematosus (3) and peripheral vascular disease (5).

The immunological assessment of these chronically infected patients showed evidence of defective cell-mediated immunity. For instance, there was absence of delayed-type hypersensitivity to trichophytin and poor lymphocyte transformation to dermatophyte antigens[15,16], but not T-lymphocyte mitogens such as phytohaemagglutinin. This contrasted with enhanced lymphocyte transformation in patients with *T. interdigitale* infections. However, similar T-lymphocyte hyporeactivity was also seen in patients with 'dry-type' *T. rubrum* infections who had responded to treatment. Therefore it would appear that *T. rubrum* infections of the moccasin type are associated *per se* with these poor responses. While other workers have shown similar results, one study in Denmark has demonstrated adequate cell-mediated immune responses in patients with acute foot infections caused by *T. rubrum*[17].

About 40–60% of patients with chronic *T. rubrum* infections of the feet show immediate-type hypersensitivity to the organism, an observation which may correlate with the high proportion of atopics with this form of infection. There is also a significant increase in the proportion of atopics with foot infections due to *T. rubrum* in community studies of school children[18] who may be exposed to infection. Whether the association between atopy and chronic *T. rubrum* infection is based on the T-lymphocyte dysfunction seen in some atopics, or another factor such as enhanced fungal adherence, is not known.

CHRONIC TINEA CORPORIS

Chronic tinea corporis is not common in Europe and the USA. In the UK survey mentioned previously it accounted for less than 5% of treatment-unresponsive dermatophytosis and in all cases was caused by *T. rubrum* (Fig. 2.2). Other organisms including *T. violaceum* or *T.*

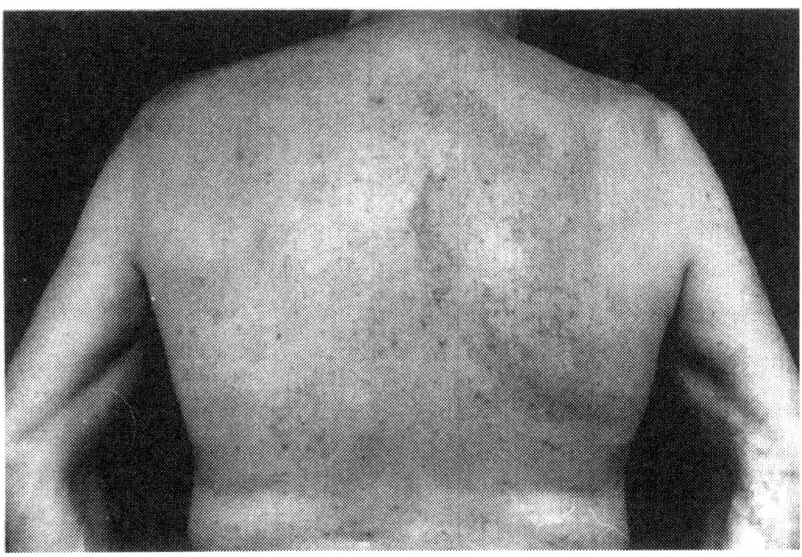

FIGURE 2.2 Chronic tinea corporis due to *T. rubrum*

tonsurans may produce similar problems on occasions, and although it is geographically restricted, the form of tinea corporis which is most often associated with chronicity and relapse is tinea imbricata.

Tinea imbricata is an anthropophilic infection mainly confined to remote tropical areas of Southeast Asia, the West Pacific and South

America. The disease is caused by *T. concentricum*, and infected patients show characteristic concentric rings of scales which may cover the greater part of the body surface (Fig. 2.3). There is almost no spontaneous recovery, many patients having the infection for years from early infancy if they remain in an endemic area. Relapse or

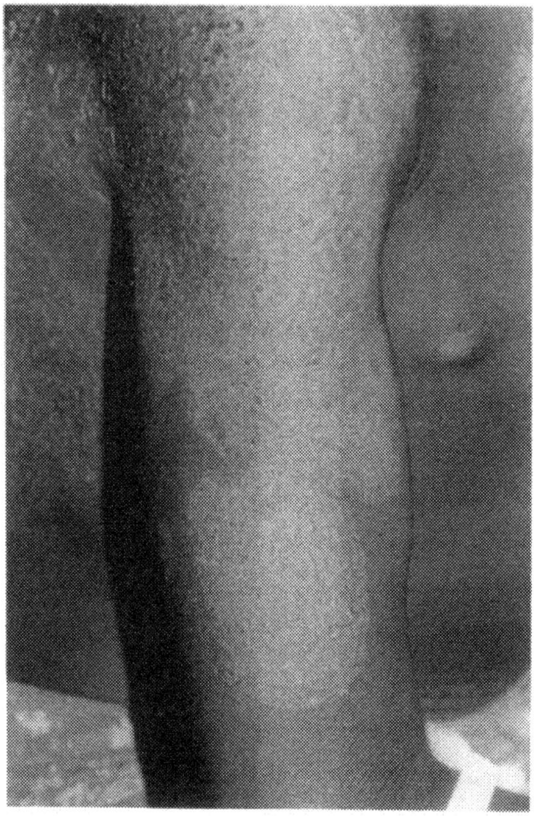

FIGURE 2.3 Tinea imbricata

reinfection is usual after treatment if the patient does not leave the endemic zone.

A number of intriguing factors have emerged about this disease. In the first instance it appears from a survey carried out in northern Papua New Guinea that in some areas the incidence of cases is compatible with the proposition that susceptibility to the infection is governed by an autosomal recessive gene[19]. While this is difficult to prove, population genetic studies point strongly to such a mechanism. Secondly, patients with *T. concentricum* infection show similar evidence of poor T-lymphocyte-mediated responses to the specific organism in skin tests and leukocyte migration inhibition studies, as patients with 'dry type' *T. rubrum* infections[20]. Likewise they show a high level of immediate-type skin reactions to trichophytin; unlike those with 'dry-type' *T. rubrum* infections they have high titres of circulating specific antibody of all classes. In these patients it is not possible to define the presence of atopy on clinical grounds because both hay fever and asthma are uncommon in endemic areas. However, it is possible that infections have occurred predominantly but not exclusively in those who are genetically predisposed to high IgE production, and again this may interfere with the development of T-lymphocyte reactivity.

The consequences of this apparent heightened susceptibility and lack of immunity are that patients cleared of infection by treatment with griseofulvin are still at risk from reinfection. It has been shown that considerable quantities of viable *T. concentricum* can be isolated from the houses of infected patients, and unless their immediate environment as well as personal contacts are fungus-free, reinfection seems inevitable[21].

CONCLUSION

An attempt has been made to explain the persistence of some dermatophyte infections in terms of their location and the type of immune response seen in the host. It is also clear that the nature of the infecting organism is also important with anthropophilic organisms being most associated with persistence – a factor of presumed evolutionary importance. Treatment failure in the face of these obstacles can often be explained by lack of drug penetration into relevant areas such as nails and failure of the host's immune response when the drug is only achieving fungistatic levels in stratum corneum. Relapse will occur if exposure is continued and if little immunity has developed as a consequence of the initial infection.

In addition to these factors, the behaviour of the drug in an individual patient may also be important. For instance poor absorption of griseofulvin may lead to inadequate tissue levels, and although this may occur without any apparent explanation, it may also follow competition by other drugs such as phenobarbitone[22]. Likewise it has become apparent that some patients do not absorb ketoconazole well at normal therapeutic doses. Drug resistance has also been cited as a problem. In a prolonged study of griseofulvin in onychomycosis Davies and co-workers[23] concluded that there were no significant changes in the minimum inhibitory concentration values of griseofulvin for the organism isolated before as compared to during treatment[13]. However, this view has been challenged more recently[24], when it was found that patients judged to be griseofulvin treatment failures were significantly more likely to have organisms with high MIC (minimum inhibitory concentration) values; this was particularly true of tinea corporis. The mechanism of this 'resistance' is unknown, but it has been shown that dermatophytes can be 'trained' to grow in increasing concentrations of griseofulvin *in vitro* and it is possible that a similar adaptation is involved here. Imidazole resistance has been reported for *Candida albicans* in patients with chronic mucocutaneous candidosis[25] receiving ketoconazole, but similar findings have not been shown in dermatophytosis. Once again the mechanism is unknown.

The implications for the clinician of the existence of refractory dermatophyte infections can be summarized simply. Certain dermatophyte infections, including most of those cited here, have a low spontaneous recovery rate and it seems sensible to treat these from the outset with oral antifungal drugs. The latter should be given in adequate doses, e.g. griseofulvin 10–15 mg kg^{-1} daily and for adequate lengths of time. It is important to try to confirm mycological recovery by scraping and culture before stopping treatment. In those who fail to respond it is worth checking the MIC of the organism and serum levels of drugs. Where relapse is a common hazard it is also useful to treat other family members who are affected, and in cases of anthropophilic scalp ringworm the source of infection should be identified if possible.

Despite these measures treatment failures are inevitable. It is useful in these circumstances to consider changing the drug from griseofulvin to ketoconazole or vice-versa. Given the rare but potentially serious hazard of hepatotoxicity if ketoconazole is used, patients should have their bliver function monitored regularly and be asked to report any relevant side-effects immediately[26]. Onychomycosis presents perhaps the greatest challenge, but here it may be possible to use additional therapy, particularly if the proximal margin of infection is well clear of the nail fold. These measures include nail removal using 40% urea[27], partial surgical ablation of the nail[28] or topical 28% tioconazole solution[29].

For those patients who remain infected there are no easy answers. However, there are three new oral antifungals with potential promise in dermatophytoses – fluconazole, itraconazole and terbinafine. Fluconazole is a triazole which is considerably better absorbed after oral administration than the others[30].

Itraconazole shows good *in vitro* activity against mould fungi including dermatophytes, and early clinical studies have shown clinical efficacy in resistant ringworm infections[31]. Terbinafine is an allylamine, a novel group of drugs with a mode of action which is different from the imidazoles. It inhibits the formation of squalene epoxide in the biosynthesis of the cell membrane[32]. While the assessment of all three of these drugs is at an early stage, it is possible that one of these drugs may provide a more satisfactory answer to chronic dermatophyte infections.

REFERENCES

1. Roberts, S. O. B and MacKenzie, D. W. R. (1979). Mycoses. In Rook, A., Wilkinson, D. S. and Ebling, F. J. G. (eds). *Textbook of Dermatology*. (Oxford: Blackwell Scientific Publications)
2. Svejgaard, E. (1985). Immunologic investigations of dermatophytes and dermatophytosis. *Seminars in Dermatology – Mycology*. Vol. 4, pp. 201–221. (New York: Thieme)
3. Kaaman, T. (1985). Dermatophyte antigens and cell-mediated immunity in dermatophytosis. In *Current Topics in Medical Mycology*. pp. 117–134. (New York: Springer-Verlag)
4. Hay, R. J. (1985). Superficial fungal diseases. In Stone, J. (ed.). *Dermatologic Immunology and Allergy*. pp. 501–510. (St Louis: C. V. Mosby)
5. Davies, R. R. and Zaini, F. (1984). Trichophyton rubrum and the chemotaxis of polymorphonuclear leucocytes. *J. Med. Vet. Mycol.*, **22,** 65–71
6. Braathen, L. R. and Kaaman, T. (1983). Human epidermal Langerhans cells induce cellular immune responses to trichophytin in dermatophytosis. *Br. J. Dermatol.*, **109,** 295–300
7. Grappel, S. F., Blank, F. and Bishop, C. T. (1972). Circulating antibodies in dermatophytosis. *Dermatologica,* **144,** 1–11

8. Calderon, R. A. and Hay, R. J. (1984). Cell-mediated immunity in experimental murine dermatophytosis. II: Adoptive transfer of immunity to dermatophyte infection by lymphoid cells from donors with acute or chronic infections. *Immunology*, **53**, 465–472
9. Jones, H. E., Reinhardt, J. H. and Rinaldi, M. G. (1974). Acquired immunity to dermatophytes. *Arch. Dermatol.*, **109**, 840–848
10. Rasmussen, J. E. and Ahmed, A. R. (1978). Trichophytin reactions in children with tinea capitis. *Arch. Dermatol.*, **114**, 271–273
11. Zaias, N. (1972). Onychomycosis. *Arch. Dermatol.*, **105**, 263–274
12. Hay, R. J. and Baran, R. (1984). Fungal (onychomycosis) and other infections of the nail apparatus. In Baran, R. and Dawber, R. P. (eds). *Diseases of the Nails and their Management*. pp. 121–144. (Oxford: Blackwell Scientific Publications)
13. Rippon, J. W. (1985). The changing epidemiology and emerging patterns of dermatophyte species. In: *Current Topics in Medical Mycology*, pp. 208–234. (New York: Springer-Verlag)
14. Hay, R. J. (1982). Chronic dermatophyte infections. I: Clinical and mycological features. *Br. J. Dermatol.*, **106**, 1–7
15. Hanifin, J. M., Ray, L. F. and Lobitz, W. C. (1974). Immunological reactivity in dermatophytosis. *Br. J. Dermatol.*, **90**, 1–8
16. Hay, R. J. and Shennan, G. (1982). Chronic dermatophyte infections. II: Antibody and Cell-mediated Immune Responses. *Br. J. Dermatol.*, **106**, 191–198
17. Svejgaard, E., Thomsen, M., Morling, N. and Christiansen, A. H. (1976). Lymphocyte transformation *in vitro* in dermatophytosis. *Acta Pathol. Microbiol. Scand.*, **84**, 511–519
18. Svejgaard, E., Albrectsen, B. and Baastrup, N. (1983). The occurrence of tinea of the feet in 15 year old school children. *Mykosen*, **26**, 450–454
19. Serjeantson, S. and Lawrence, G. (1977). Autosomal recessive inheritance of susceptibility to tinea imbricata. *Lancet*, **1**, 13–15
20. Hay, R. J., Reed, S., Talwat, E. and MacNamara, K. (1983). Immune responses of patients with tinea imbricata. *Br. J. Dermatol.*, **108**, 581–586
21. DeVroey, C. (1985). Epidemiology of ringworm. *Seminars in Dermatology*. Vol. 4, pp. 185–200. (New York: Thieme)
22. Davies, R. R. (1980). Griseofulvin. In Speller, D. C. E. (ed.). *Antifungal Chemotherapy*. pp. 149–182. (Chichester: Wiley)
23. Davies, R. R., Everall, J. D. and Hamilton, E. (1967). Mycological and clinical evaluation of griseofulvin for chronic onychomycosis. *Br. Med. J.*, **iii**, 464–469
24. Artis, W. M., Odle, B. M. and Jones, H. E. (1981). Griseofulvin-resistant dermatophytosis correlates with *in vitro* resistance. *Arch. Dermatol.*, **117**, 16–21
25. Ryley, J. F., Wilson, R. G. and Barrett-Bee, K. J. (1984). Azole resistance in *Candida albicans*. *J. Med. Vet. Mycol.*, **22**, 53–63
26. Lewis, J. H., Zimmerman, J. H., Benson, G. D. and Ishak, K. G. (1984). Hepatic injury associated with ketoconazole therapy – an analysis of 33 cases. *Gastroenterology*, **86**, 503–512
27. White, M. I. and Clayton, Y. M. (1982). The treatment of fungus and yeast infections of nails by the method of chemical removal. *Clin. Exp. Dermatol*, **7**, 273–276
28. Baran, R. and Hay, R. J. (1985). Partial surgical avulsion of the nail in onychomycosis. *Clin. Exp. Dermatol.*, **10**, 413–418
29. Hay, R. J., Mackie, R. M. and Clayton, Y. M. (1985). Tioconazole nail solution –

an open study of its efficacy in onychomycosis. *Clin. Exp. Dermatol.*, **10,** 111–115
30. Richardson, K., Brammer, K.W., Marriott, M.S. and Troke, P.F. (1985). Activity of UK-49,858, a Bis-triazole derivative, against experimental infections with *Candida albicans* and *Trichophyton mentagrophytes*. *Antimicrob. Ag. Chemother.*, **27,** 832–835
31. Hay, R.J. and Clayton, Y.M. (1986). The treatment of chronic dermatophyte infections and chronic oral candidosis with itraconazole. *Rev. Infect. Dis.* (In press)
32. Ryder, N.S. (1985). Effect of allylamine antimycotic agents on fungal sterol biosynthesis measured by sterol side-chain methylation. *J. Gen. Microbiol.*, **131,** 1595–1602

3

PARONYCHIA

DAVID T. ROBERTS

PARONYCHIA

Paronychia exists in both acute and chronic forms, but they are essentially different diseases with a separate aetiology.

Acute paronychia

This condition presents as a painful swelling of the posterior or lateral nail fold. Untreated it will usually point close to the nail and is easily dealt with by incision, thus allowing the pus to drain. This can be done without anaesthetic but more commonly a short-acting freezing agent such as ethyl chloride or even liquid nitrogen is used. This form of paronychia is usually due to staphylococcal infection, often secondary to minor local trauma, although a positive history of such trauma is not always obtained. Cases presenting with painful swelling but no obvious pointing should be treated with an antibiotic with antistaphylococcal activity. This sometimes renders a drainage procedure unnecessary.

Chronic paronychia

Chronic paronychia is commonly seen in dermatological practice. It is one of those conditions which often appears trivial to the observer, but is painful and distressing to the patient who becomes even more distressed when confronted by the apparent indifference of the doctor.

Successful treatment depends upon the aetiology, and it is convenient to classify chronic paronychia as either primary or secondary.

Primary chronic paronychia

This condition presents with painful swelling of the posterior and lateral nail folds which become erythematous and shiny-looking. The cuticle is usually seen to be detached from the nail plate, and this is an important diagnostic feature (Fig. 3.1). In the early stages the nail

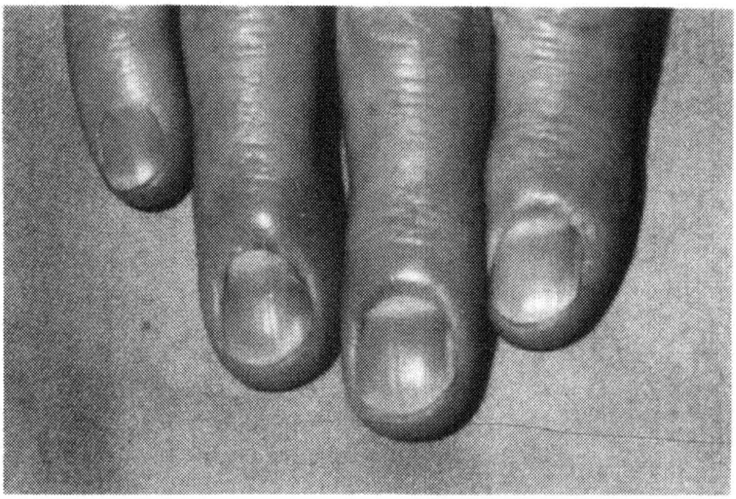

FIGURE 3.1 Chronic paronychia showing swelling and inflammation of nail folds

itself is normal, but as the disease progresses the proximal and lateral areas of the nail plate become discoloured and eventually dystrophic (Fig. 3.2). The disease is much more often seen in females, usually those between 30 and 60 years of age[1], and is thought to be causally related to chronic immersion of the hands in water. Intercurrent diseases such as diabetes or Raynaud's phenomenon[2] increase the risk, as do occupational factors which lead to long-standing immersion of the hands. It is therefore commonest in housewives, barmaids, nurses, hairdressers and catering workers. Males with similarly 'wet' occupations are also at risk[3]. In children it is relatively rare but is some-

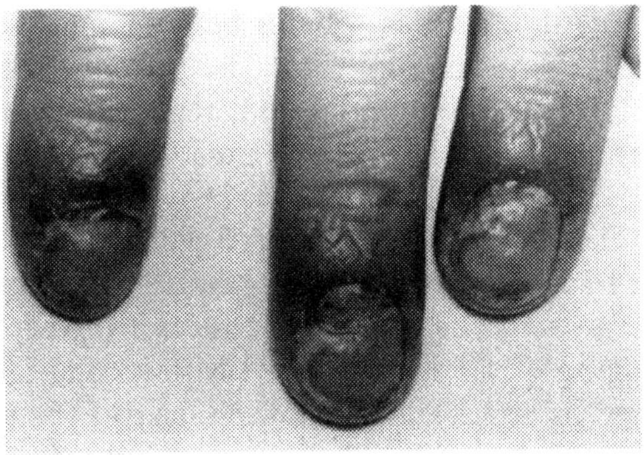

FIGURE 3.2 Chronic paronychia showing dystrophy of the proximal nail plate

times seen secondary to thumb-sucking[4] (Fig. 3.3). It has been said that the disease affects the middle and right index fingers more often[3]. This may be statistically so but should not be used as an important diagnostic criterion as all fingers may be affected. However, the above-mentioned digits are reckoned to be more prone to trauma during activities such as washing clothes.

Chronic low-grade inflammation continues in the nail folds and is

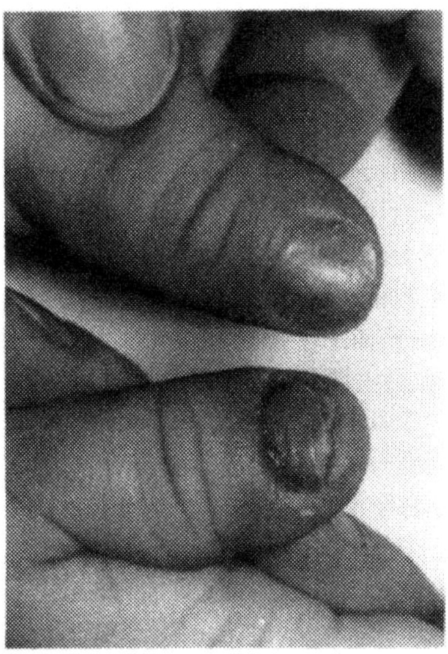

FIGURE 3.3 Chronic paronychia secondary to thumb sucking

punctuated by acute episodes when a bead of pus can often be expressed by pressing on the nail fold. The pus is said to form in the pocket under the nail fold and does not represent abscess formation within the perionychium[5]. Repeated episodes of acute on chronic inflammation lead eventually to yellow, brown or black discoloration of the proximal lateral areas of the nail plate. This discoloration is said to be due to dihydroxyacetone produced by organisms found in the nail fold[4]. *Pseudomonas* specifically results in a green discoloration. Transverse ridging of the nail, coupled with a rough friable surface, follows repeated infection, and the overall size of the nail is reduced. Eventually the condition becomes cosmetically as well as symptomatically distressing. It is generally accepted that a number of organisms coexist in the nail fold in cases of primary chronic paronychia. These include yeasts of the species *Candida* and a wide variety of bacteria. In a review of 590 cases, *Candida* was found in 70%[3] and an incidence of 97% has been reported in a smaller number of cases[6]. This latter work also revealed that various bacteria were present in 29 out of 34 cases. *Escherichia coli* was predominant, followed by *Proteus vulgaris,* diphtheroids, coagulase-negative staphylococci, *Klebsiella aerogenes, Pseudomonas pyocyanea,* and other varieties of enterococci. It would appear that the pattern of organisms isolated varies with the chronicity of the disease. Specimens taken from very early lesions present only for a day or two before a pocket developed under the nail fold, revealed only *Staph. aureus*[7]. Later lesions from the same patients tended to grow *Candida* spp and in this series *C. parapsilosis, C. tropicalis, C. krusei* and *C. guilliermondi* were isolated although, of course, *C. albicans* was most commonly found. A later comparative study[8] looked at the bacteriology and mycology of 91 infected nail folds in paronychia patients and compared the results with those obtained from 72 normal control nails. In the paronychia group *C. albicans* was isolated from 77 nail folds along with *S. pyogenes* (44 cases), coliform spp (66 cases), *Proteus* spp. (48 cases), *Ps. aeruginosa* (pyocyanea) (34 cases), and *Strept. faecalis* (25 cases). In the control group of normal nails only *Staph. epidermidis, Staph. pyogenes* and *Strept. faecalis* were found 57, 18 and 11 cases respectively. In this series *Candida* spp. did not appear, therefore, to be commensals of normal nail folds.

The studies confirm the presence of both bacteria and yeasts in infected nail folds and the varying incidence of recovery of organisms may depend upon the methods used for obtaining specimens. In one study[6] an orange stick was pushed into the pocket under the nail fold and the debris scraped out and immediately plated, whilst in another[8] a platinum loop containing a drop of saline was passed into the nail fold pocket. Neither one of these methods is likely to endear the doctor to the patient, but it is clearly important to obtain material from beneath the nail fold. This can be achieved reasonably satisfactorily by soaking a specimen swab in saline and passing it along the proximal nail fold, ensuring that as much saline as possible washes under the fold into the pocket. The resulting contents of the swab are then cultured for fungi and bacteria.

The work quoted reveals the nature of the flora in cases of paronychia but does little to indicate whether *Candida* spp. or bacteria are the predominant pathogens at which treatment should be aimed, and this remains a controversial issue. It is well recognized that other foreign material such as wax[9] and hair[10], as well as foodstuffs, may collect under the nail fold and contribute to the chronicity of the disease by causing retraction of the nail fold. Others believe that completely different factors such as immediate or delayed hypersensitivity to chemicals contained in food are in part responsible for the disease process[11]. Attempts have been made to reproduce the disease by introducing *Candida* into the nail fold. In one such instance[12] both middle finger nail folds, initially normal, were macerated by occlusion with waterproof tape prior to the introduction of *C. albicans*. However, both nail folds became colonized with *C. guilliermondi* and later by *C. parapsilosis*. The nail folds were kept occluded but only on two short-lived occasions did the typical changes of swelling and tenderness seen in paronychia occur. Both these episodes were coincident with low-grade viral infections and each time the yeasts were joined in the nail fold by *Ps. pyocyanea*, which disappeared as the acute inflammation subsided. A single experiment such as this is inconclusive; merely suggesting that both yeasts and bacteria are necessary to produce the full-blown clinical appearances, and it has always been suggested that treatment be directed at both organisms. The site of infection deep in the nail fold pocket and perionychium indicates that systemic treatment is likely to be more successful than topical. Until recently no oral anticandidal agent suitable for long-term use was available. The advent of the drug ketoconazole, an orally absorbed imidazole, with a broad-spectrum activity against yeasts as well as ringworm fungi, allowed a therapeutic study against *Candida* to be carried out in patients with chronic paronychia[13]. Twelve female patients, all of whom had signs and symptoms of chronic paronychia for a mean duration of 10 years, were examined mycologically. In 10 out of 12 patients *Candida* was isolated from saline swabs taken from the nail fold. *C. albicans* was the predominant species but *C. parapsilosis* was found in two patients. Patients were given ketoconazole 200 mg daily in a single dose for a mean period of 10 months. One patient dropped out of the study after 5 months, having developed hepatitis. At the end of the treatment period 11 of the 12 had improved markedly on the basis of the clinical parameters of paronychia, discolouration, malformation and onycholysis, and although six of the patients remained mycologically positive for *Candida* the number of colonies grown from each swab was quantitatively much less. Subjectively all 11 patients were pleased with the result and claimed the drug was the most effective treatment they had received.

A study such as this adds some weight to previous opinion[14] that yeasts of the *Candida* species, usually *C. albicans*, are the predominant pathogens. However there is little doubt that bacteria mainly originating in the gut[8] play a role in chronic paronychia and may even initiate it.

The source of the organisms remains obscure, but the mouth and gut are the most likely reservoirs of infection. Now that reliable methods are available for biotyping *Candida* species the likely source ought to be able to be pinpointed by examination of appropriate specimens from mouth and gut as well as the nail folds. It also seems reasonable to postulate that there may be immunological reasons for the persistence of this condition in certain patients. Primary chronic paronychia therefore remains something of an enigma in terms of its aetiology.

Secondary chronic paronychia

Fungal nail infections due to *Hendersonula toruloidea* or *Scytalidium hyalinum* may be associated with paronychia[15]. Digital enchondromata[16] and other tumours and cysts are unusual causes of paronychia (Fig. 3.4), as is radiodermatitis. Tungiasis, an infection caused by the burrowing flea *Tunga penetrans*, presents as paronychia of the

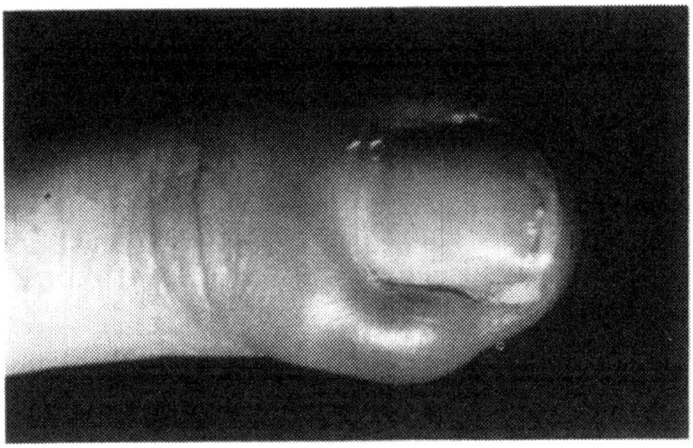

FIGURE 3.4 Chronic paronychia secondary to mucous cyst (patient of Dr Alan Lyell)

toe nails[17]. Synthetic retinoids are increasingly used in the treatment of a number of hyperkeratotic and proliferative skin disorders, notably psoriasis, and a number of nail abnormalities including paronychia have been reported following their use[18,19] (Fig. 3.5).

A number of diseases such as syphilis, pemphigus and parakeratosis pustulosa produce lesions adjacent to the nails, which may mimic paronychia.

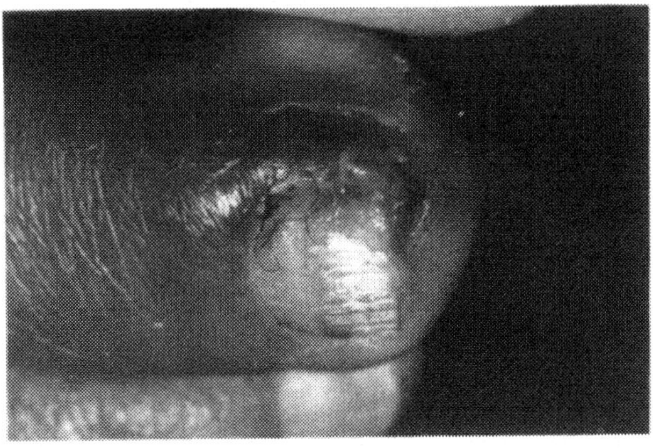

FIGURE 3 5 Chronic paronychia secondary to etretinate therapy

Skin diseases which affect the fingers often cause swelling of the skin of the nail folds leading to detachment of the cuticle. A secondary paronychia thus follows by a similar mechanism to primary chronic paronychia in that the nail fold pocket becomes colonized with bacteria and yeasts. Psoriasis and dermatitis are common examples of such diseases. Similarly a variety of artefact can produce a chronic paronychia (Fig. 3.6). In such cases the cuticle is detached by the patients themselves and telltale central cross-ridging of the nail is seen.

Conditions which, for various reasons, interfere with host defence mechanisms may also present as paronychia and diabetes, Raynaud's disease and chronic mucocutaneous candidiasis come into this category.

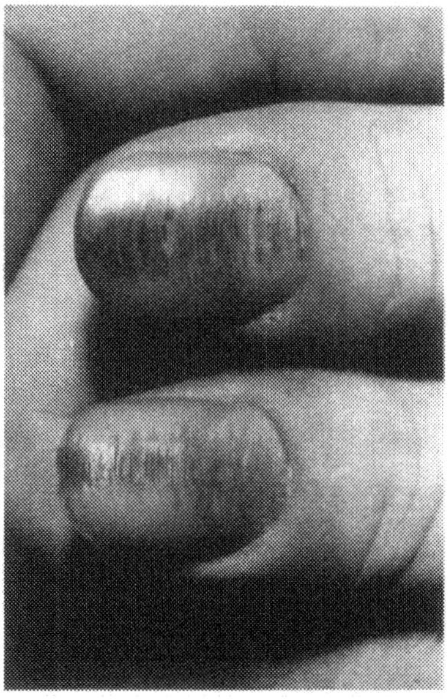

FIGURE 3.6 Artefact as a cause of chronic paronychia (patient of Dr Alan Lyell)

Treatment of chronic paronychia

The treatment of primary chronic paronychia is difficult and tedious, and requires considerable patience on the part of both doctor and patient. Topical antifungal preparations active against *Candida* may be helpful and currently imidazole lotions such as clotrimazole or econazole are the most appropriate agents because, in addition to their anticandidal action, they may have slightly greater antibacterial activity than does nystatin. Such preparations should be applied as drops to the area of the posterior and lateral nail folds twice daily, and rubbed in proximally so as to force as much of the lotion as possible into the nail fold pocket. Alternatively, if there is evidence that bacterial infection is consistently present, the time-honoured remedy of sulphacetamide paint 15% in 50% spirit[14] can be applied in a similar fashion. This preparation is simply prepared by diluting 30% sulphacetamide eye drops (Albucid eye drops BNF) in 50% spirit. Other antiseptic agents advocated include Burow's solution (1:40) and 4% thymol in chloroform[20]. Anticandidal and antibacterial agents used alternatively night and morning have been advocated[8], and sometimes a systemic antibiotic is required for acute flare-ups.

Systemic anticandidal therapy in the form of ketoconazole is likely to be helpful[13] (Fig. 3.7) but reports of hepatotoxicity[21] are worrying

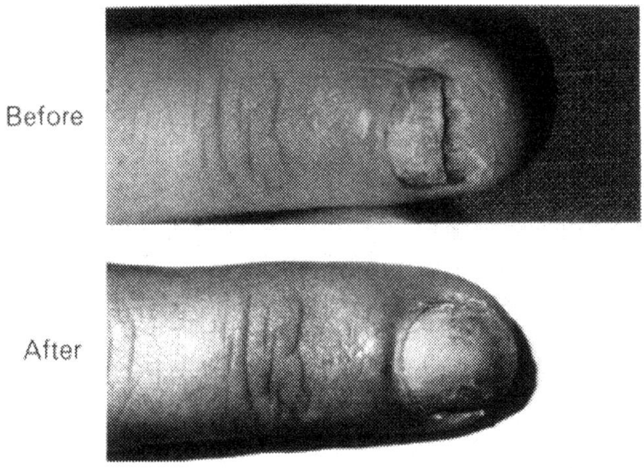

FIGURE 3.7 Chronic paronychia with nail dystrophy before and after treatment with ketoconazole

when treating a totally non-life-threatening disease. However, its use may sometimes be justified when topical treatment has failed and the patient remains subjectively distressed by the symptoms. It is the treatment of choice in chronic mucocutaneous candidiasis[22]. If there is overt evidence of candidal infection in other skin sites, or in the mouth or vagina, then the appropriate therapy should be instituted.

Patients must be instructed and constantly encouraged to keep their hands as dry as possible by wearing rubber gloves, preferably with a pair of thin cotton gloves worn underneath the rubber gloves. This is not always easy when the problem is occupational, or in mothers with young babies. In cases of secondary chronic paronychia the underlying or intercurrent disease should obviously be treated along with the paronychia.

Whichever method of treatment is chosen, it must be continued until the inflammatory swelling resolves and the cuticle becomes re-attached to the nail plate. This is likely to take several months, and treatment may need to be continued for a year or more if the nail itself is dystrophic and harbours the organism.

REFERENCES

1. Esteves, J. (1959). Pathogenesis and treatment of chronic paronychia. *Dermatologica,* **119,** 229–233
2. Hellier, F. (1955). Chronic perionychia: aetiology and treatment. *Br. Med. J.,* **ii,** 1358–1360
3. Frain-Bell, W. (1957). Chronic paronychia. *Trans. St John's Hosp. Dermatol. Soc.,* **38,** 29–33
4. Stone, O.J. and Mullins, J.F. (1968). Chronic paronychia in children. *Clin. Paediat.,* **7,** 104–107
5. Hay, R.J. and Baran, R. (1984). Fungal (Onychomycosis) and other infections of the nail apparatus. In Baran, R. and Dawber, R.P.R. (eds) *Diseases of the Nails and their Management.* (Oxford: Blackwell Scientific Publications)
6. Marten, R.H. (1959). Chronic paronychia. A mycological and bacteriological study. *Br. J. Dermatol.,* **71,** 422–426
7. Whittle, C.H., Moffat, J.L. and Davis, R.A. (1959). Paronychia or perionychia: aetiological aspects. *Br. J. Dermatol.,* **71,** 1–11
8. Barlow, A.J.E., Chattaway, F.W., Holgate, M.C. and Aldersly, T. (1970). Chronic paronychia. *Br. J. Dermatol.,* **82,** 448–453
9. Stone, O.J., Mullins, J.F. and Head, E.S. (1964). Chronic paronychia. Occupational material. *Arch. Environ. Hlth,* **9,** 585–588
10. Stone, O.J. (1975). Chronic paronychia in which hair was a foreign body. *Int. J. Dermatol.,* **9,** 661–663
11. Zaias, N. (1980). *The Nail in Health and Disease.* (New York: Spectrum Publications)
12. Whittle, C.H. and Gresham, G.A. (1963). Paronychia experimentally induced. *J. Invest. Dermatol.,* **40,** 267–269
13. Roberts, D.T. and Gentles, J.C. (1985). The treatment of chronic paronychia.

Presented at the *IXth International Society of Human and Animal Mycology*. 19–24 May Atlanta, Georgia
14. Samman, P. D. (1972). The nails. In Rook, A., Wilkinson, D. S. and Ebling, F. J. G. (eds). *Textbook of Dermatology,* 2nd Edn, pp. 1642–1671. (Oxford: Blackwell Scientific Publications)
15. Roberts, S. O. B. and McKenzie, D. W. R. (1979). Mycology. In Rook, A., Wilkinson, D. S. and Ebbing, F. J. G. (Eds). *Textbook of Dermatology,* 3rd Edn, pp. 767–868. (Oxford: Blackwell Scientific Publications)
16. Shelley, W. B. and Ralston, E. L. (1964). Paronychia due to enchondroma. *Arch. Dermatol.,* **90,** 412–413
17. Zalar, G. L. and Walter, R. P. (1980). Infestation by *Tunga penetrans. Arch. Dermatol.,* **116,** 80–81
18. Baran, R., Brun, P. and Juhlin, L. (1983). Leuchonychie transversale induite par etretinate. *Ann. Dermatoven,* **109,** 367–371
19. Ferguson, M. M., Simpson, N. B. and Hammersly, N. (1983). Severe dystrophy associated with retinoid therapy. *Lancet,* **2,** 974
20. Wilson, J. W. (1965). Paronychia and onycholysis, aetiology and therapy. *Arch. Dermatol.,* **92,** 726–730
21. Van Tyle, J. H. (1984). Ketoconazole, mechanism of action, spectrum of activity, pharmacokinetics, drug interactions, adverse reactions and therapeutic use. *Pharmacotherapy,* **4,** 343–373
22. Hay, R. J. (1981). Management of chronic mucocutaneous candidosis. *Clin. Exp. Dermatol.,* **6,** 515–519

4

PITYRIASIS VERSICOLOR

S. O. B. ROBERTS

Pityriasis versicolor was first recognized as a fungal infection by Eichstedt in 1846. Now 140 years on it is known to represent the human host's response to multiplication and minimal invasion by an organism normally resident on the skin. That organism, variously known as *Malassezia furfur* and *Pityrosporum orbiculare*, is a lipophilic yeast which characteristically undergoes mycelial transformation as it assumes its minor pathogenic role. In the past 20 years pityriasis versicolor research has involved the study of the normal yeast flora of the skin, yeast/mycelial transformation and the host factors which lead to the shift in the host/parasite relationship and result in this common but fascinating disorder.

NORMAL FLORA

The light microscopists of the late nineteenth century were familiar with oval and round yeast-like structures in scrapings from scaly scalps of patients with, for example, psoriasis, seborrhoeic dermatitis and pityriasis capitis[1]. The initial suggestion that they had a causal role in these conditions came to be rejected as the oval yeast – by this time known as *Pityrosporum ovale* – was recognized as a regular inhabitant on the skin of the healthy scalp. The idea that a particular micro-organism should exist as a harmless commensal at one time but become a pathogen at another was not yet appreciated. In Eichstedt's day other fungal pathogens were becoming recognized and isolated in culture, e.g. *Trichophyton schoenleinii* from favus. *Malassezia furfur*, as it became known, was clearly visible in scrapings as round 'spores' and filaments but could not be cultured, nor was it recognized as having any similarity to the yeasts seen by others in scalp scrapings.

By the early 1960s when Mary Marples was compiling her classic work *The Ecology of the Human Skin*[2] it was generally accepted that *P. ovale* was a member of the resident skin flora of the scalp, face and other sites and could be exonerated as a pathogen in psoriasis or seborrhoeic dermatitis. At that time it was known that *P. ovale* was lipophilic and lipid-dependent. Moreover Maurice Gordon's achievement in 1951 of isolating from the lesions of pityriasis versicolor a spherical, lipophilic yeast which he called *P. orbiculare*[3] was accepted and had been confirmed by Burke[4]. These workers believed, as did Marples, that *P. orbiculare* was the yeast phase of *M. furfur*, and they suspected on the basis of a few isolations from normal skin that it would turn out to be a resident organism. However, their inability to persuade *P. orbiculare* to undergo yeast/mycelial transformation *in vitro* and the absence of normal skin surveys left these matters open to some doubt. Since then the normal skin flora has been studied extensively. Using scrapings stained with the Parker Quink/KOH technique spherical yeasts, thick-walled and budding from a narrow base, have been observed in 98% of normal adults, often in large numbers, on the chest and back[5]. Culture using olive oil over malt extract agar, yielded *P. orbiculare* regularly from these sites. *P. ovale*'s presence on scalp, face and chest has been repeatedly confirmed. The surface biopsy technique of Marks and Dawber has been used to show how these organisms cluster in the fundibular regions of hair follicles and in crevices between keratinocytes while electron microscopy has amplified our knowledge of the precise location of this normal mycoflora[6].

There are now grounds for believing that the skin is colonized by *Pityrosporum* species in the first few days of life but that numbers of yeasts are low except in the scales of cradle cap, until later childhood. They probably rise to adult levels by the mid-teens. Certainly Faergemann and Fredriksson could culture *P. orbiculare* with ease at 15 years, having failed to isolate it in infancy[7]. The apparent increase in the numbers of yeasts seems to follow the known increase in sebaceous activity at puberty but no quantitative studies of yeasts and sebum levels have been done to prove this.

In most subjects the spherical yeast *P. orbiculare* predominates on the chest (Fig. 4.1), back and upper abdomen, while the scalp (Fig.

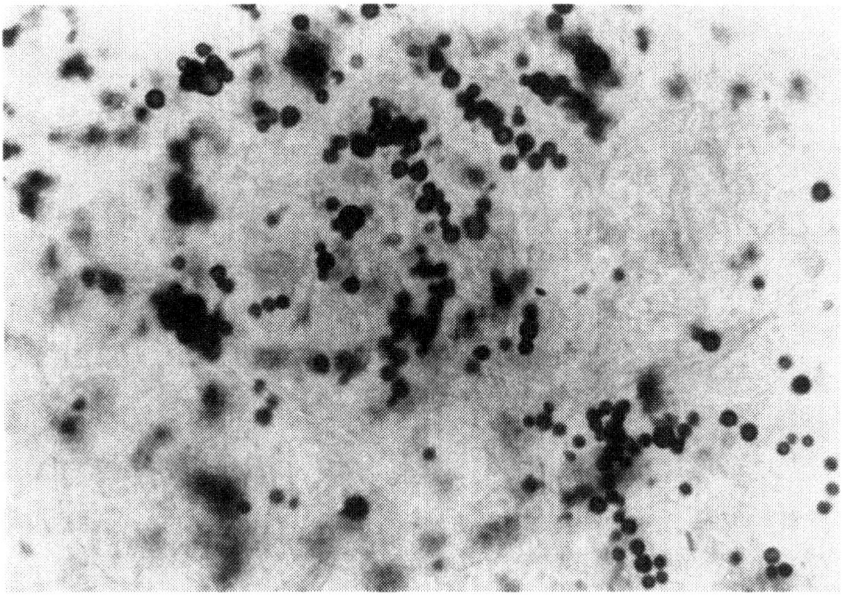

FIGURE 4.1 *P. orbiculare* in scrapings from normal skin of the chest (Parker Quink/KOH)

4.2) and face carry a population in which the oval yeast *P. ovale* is dominant. Both forms are, however, widely scattered on the body surface but simple microscopic assessment suggests they are fewer in number on the limbs and extremities. The proximal flexures seem to carry high numbers, particularly of the oval forms. More quantitative measurements are still required but McGinley's group has attempted this work using haemocytometer counts on suspensions of yeasts in skin washings obtained by the Williamson open cup technique[8]. Their figure of approximately 10 000 round yeasts/cm^2 on the normal skin of the trunk is a realistic one. On the face and scalp higher levels of yeasts, in this case *P. ovale*, have been found by the same workers[9] – 300 000/cm^2.

In the scaly lesions of pityriasis versicolor the causative fungus is found as yeasts and hyphae (Fig. 4.3). The latter are typically fragmented into short filaments. Filaments identical to these, but in smaller

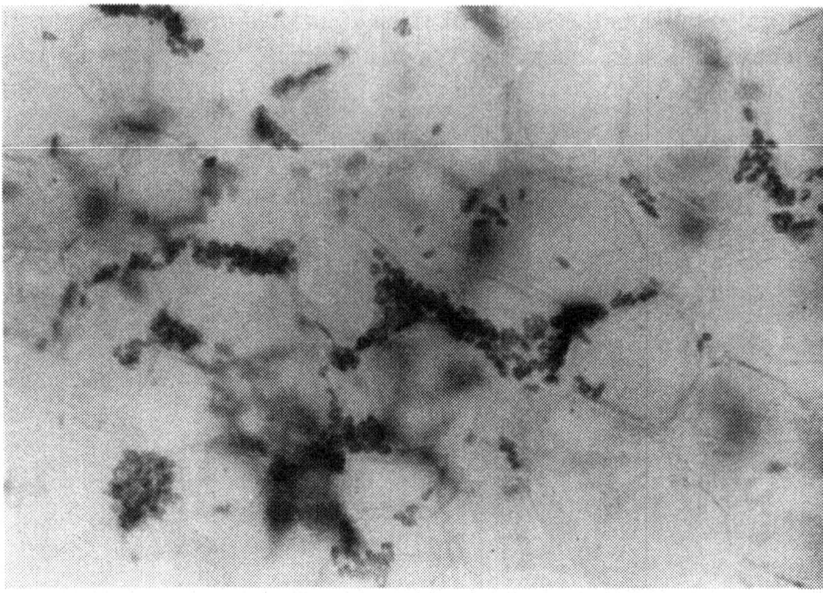

FIGURE 4.2 *P. ovale* in scrapings from normal skin of scalp (Parker Quink/KOH)

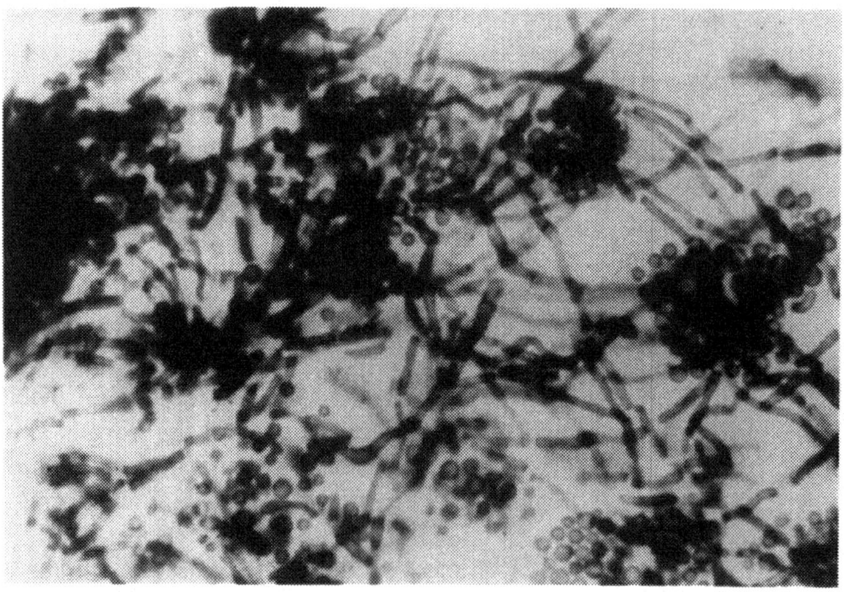

FIGURE 4.3 *M. furfur/P. orbiculare* – round yeasts and filaments in scrapings from pityriasis versicolor (Parker Quink/KOH)

numbers, may be seen in scrapings taken from the unaffected skin of patients with pityriasis versicolor, and similar structures may be observed in scrapings taken from normal subjects. From the author's experience perhaps 10% of normal subjects show these filaments[10] (Fig. 4.4). They are also found, but more frequently, in patients treated

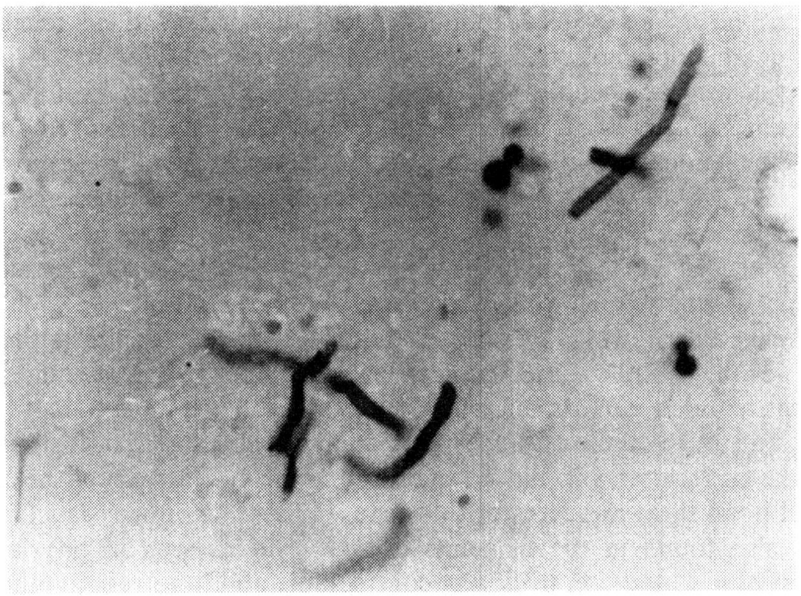

FIGURE 4.4 A few filaments and *P. orbiculare* yeasts from skin of a normal subject (Parker Quink/KOH)

with systemic steroids where they may sometimes be observed as a prelude to the development of pityriasis versicolor. They also occur in some transplant subjects who never develop this mycosis. Similar filaments observed by Di Menna in her scalp survey are reported by Marples[2], and by some of the early light microscopists who examined scalp and facial scrapings[1].

The realization that *P. ovale* required lipid for growth goes back to Panja in the late 1920s[1], and Gordon in 1951[3] demonstrated that scrapings from pityriasis versicolor would yield a spherical yeast if cultured on media overlaid with olive oil. These two *Pityrosporum* species are not the easiest of organisms to grow, and samples from skin sites showing only sparse yeasts on direct microscopy often fail to yield any organisms in culture. Solid media containing Tween's or alternatively Sabouraud's or malt extract agar overlaid with vegetable oils are generally used: the former being better for the study of colonial morphology. There is, however, no selective medium which will distinguish *P. ovale* from *P. orbiculare*. Studies of the lipid requirements of the two forms have yielded no clear results though *P. orbiculare* seems more completely lipid-dependent. The absence of reliable assimilation or fermentation tests have left microbiologists relying heavily on microscopic and colonial morphology when identifying these organisms. Their extremely thick cell walls with internal ribs are highly distinctive, so that there is little likelihood of confusing any of the *Pityrosporum* species with other yeasts. Moreover although *P. ovale* is morphologically similar to the animal species *P. pachydermatis* (also known as *P. canis*) it can easily be distinguished by its requirement for lipid, which *P. pachydermatis* does not need[1].

While the drawings of some of the early light microscopists show an oblique possibly spiral arrangement of ribs in *Pityrosporum* species it has required electron microscopy to reveal the precise nature of the cell wall and to demonstrate beyond doubt that the internal ribs begin at the bud scar and do indeed spiral round the inner aspect of the cell wall in all three members of this genus (Fig. 4.5). Similar structures are also found in the filaments of *M. furfur* as seen in pityriasis versicolor scales[11]. *P. ovale* has considerable ability to resist volume change induced by variations in osmotic pressure; a capacity attributed to its ribbed and thick cell wall[12]. This facility is obviously useful in the relatively dry inhospitable habitat of the skin surface. Neither *P. ovale* nor *P. orbiculare* seem to have any other regular habitat away from the human skin, though *P. ovale* has been isolated from the coat of the llama as well as from cats, dogs and horses. *P. orbiculare* survives reasonably well in dry conditions and the author has cultured it from skin scrapings up to 4 weeks old kept at room temperature. *P. ovale* seems to be rather more durable. It is sometimes possible to isolate *Pityrosporum* species from bed linen and clothing but the extent of its viable presence in the environment is not known. *P. pachydermatis* has occasionally been isolated from damaged human skin but does not seem to be either a regular commensal or a definite pathogen in that site, its natural habitat being the skin of a variety of animals.

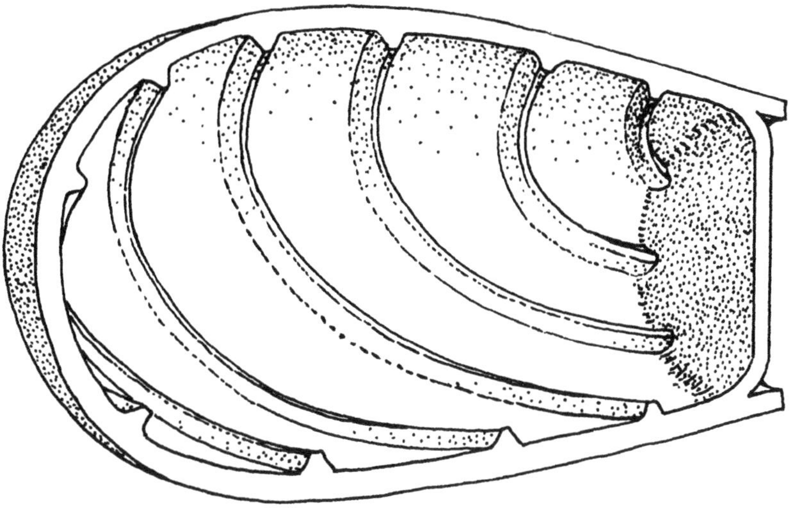

FIGURE 4.5 Diagram of *P. ovale* showing thick cell wall, spiral ribs and bud scar

In recent years Ranjandiche working in Belgium has published experiments which suggest that *P. ovale* and *P. orbiculare* are no more than morphologically distinct phases of a single species[13]. Using single spore culture he presents evidence of a change in morphology from round to oval forms as well as yeast/mycelial shift in both cases. He believes *Pityrosporum ovale* (his preferred name) is a lipophilic *Pityrosporum* species capable of a triphasic existence: as an oval yeast capable of broad-based budding, as a round yeast capable of budding from a narrow base and as mycelium. Gordon, who originally described *P. orbiculare*, has published work strongly supporting this, though he prefers the name *Malassezia furfur* to embrace *P. orbiculare* and *P. ovale* which he now considers to be one species[14]. Some other workers incline to this view but the description by Midgley (personal communication) of three distinctly recognizable stable strains of *P. ovale* – showing no sign of developing spherical forms – runs counter to the unitary position.

There is no doubt that direct microscopy of scrapings from some patients with pityriasis versicolor shows a dimorphic fungus the yeasts of which, present in clusters, are oval and associated with fragmenting mycelium entirely analogous to the *M. furfur* picture (Figs 4.6A,B).

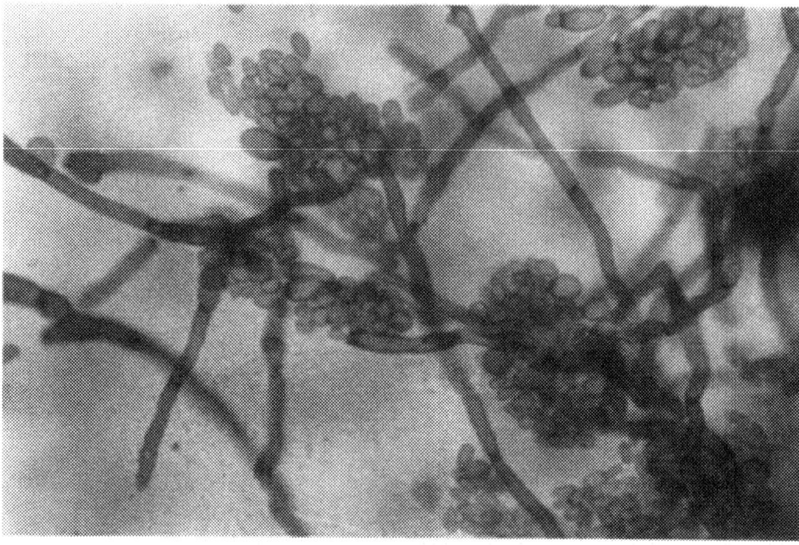

A

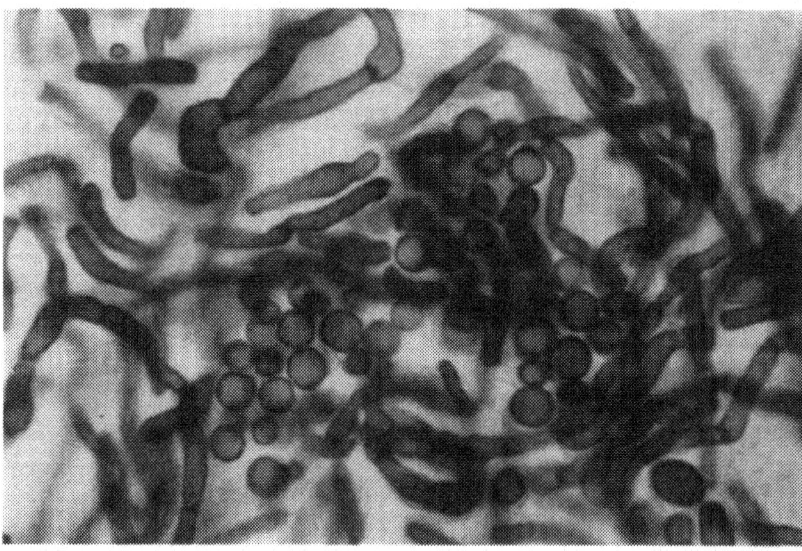

B

FIGURE 4.6 **A:** Clusters of oval yeasts and filaments in scrapings from pityriasis versicolor contrasted with **B:** classical *M. furfur* (round yeasts and filaments) (Parker Quink/KOH)

These scrapings typically yield oval yeasts in culture. The author has found this form more frequently in steroid-treated patients but has observed it in several individuals who were otherwise fit. It does not seem to be distinguishable clinically. In scaly scalps of immunosuppressed patients such an association of oval *Pityrosporum* yeasts with filaments is not uncommon. The filaments in many such cases are of a smaller diameter than is usual in classical *M. furfur*. Interestingly Vanbreuseghem, in whose laboratory Ranjandiche's work was performed, has always maintained that scrapings of pityriasis versicolor often yield oval yeasts not round ones. Many of his cultures, presumably set up in Belgium, come from samples taken in Africa from African patients and the findings on microscopy have presumably been those of classical *M. furfur*[15]. It is certainly true that a few oval yeasts may sometimes be seen in pityriasis versicolor scrapings showing predominantly classical (round form) *M. furfur*. It is possible that if cultures are only set up after a delay of some weeks *P. ovale*, though few in number, may survive at the expense of *P. orbiculare*, being apparently more durable than the round yeast. In the author's view the identity or otherwise of the two yeasts is not finally resolved. They are obviously closely related, they possess similar if not identical antigens, can both exhibit yeast/mycelial transformation and can both provoke the reaction we know as pityriasis versicolor.

PITYRIASIS VERSICOLOR – THE CLINICAL FEATURES

The classical clinical picture of pityriasis versicolor is an asymptomatic scaly eruption of the upper trunk extending often to the neck and to the proximal parts of the upper limbs (Fig. 4.7). The extension to the lower forearms, trunk, thighs and popliteal fossae are less usual but well recognized by dermatologists practising in temperate climates. However facial involvement and hand and foot lesions are distinctly uncommon and palmar ones rare (the author has seen one proven example) outside tropical zones. The scales appear as a fawn or light brown discoloration on a pale skin but as relatively paler areas on tanned or naturally dark subjects.

Depigmentation without scaling may persist after successful treatment or spontaneous resolution. In warmer climates pityriasis ver-

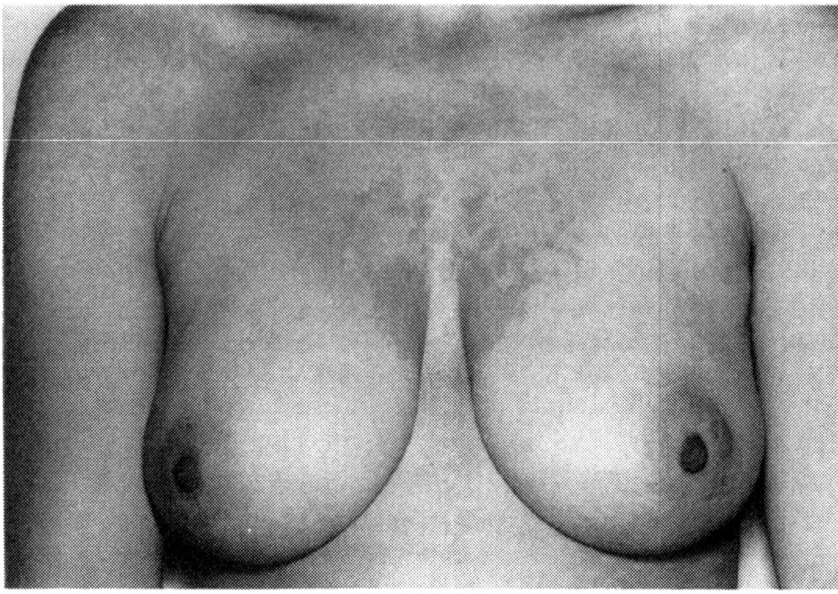

FIGURE 4.7 Pityriasis versicolor of chest and shoulders

sicolor is typically more extensive, more chronic and more liable to relapse after treatment.

Individual case reports of atypical forms of pityriasis versicolor abound: localized eruptions confined to the face, the scalp and the groins or under the straps of an army pack. The author has observed localized sub-mammary pityriasis versicolor and scaly changes restricted to the hair-bearing skin of the pubic area as well as erythrasma-like groin lesions. In some of these sites high moisture levels, so important in experimental infections, may be implicated. Certainly a particular tendency to flexural lesions seemed to be associated with increased sweating in East Anglia[5]. Very localized lesions on chest or back are quite often seen but these probably represent the initial expression of the classical eruption. Though scaling may be found in confluent sheets in some patients and as small macules (often centred on hair follicles) in others annular lesions are occasionally seen (Fig. 4.8). The latter may be particularly associated with itchy and inflamed pityriasis versicolor. However, among that minority of patients whose lesions do reveal clinically obvious inflammation most will show no central clearing.

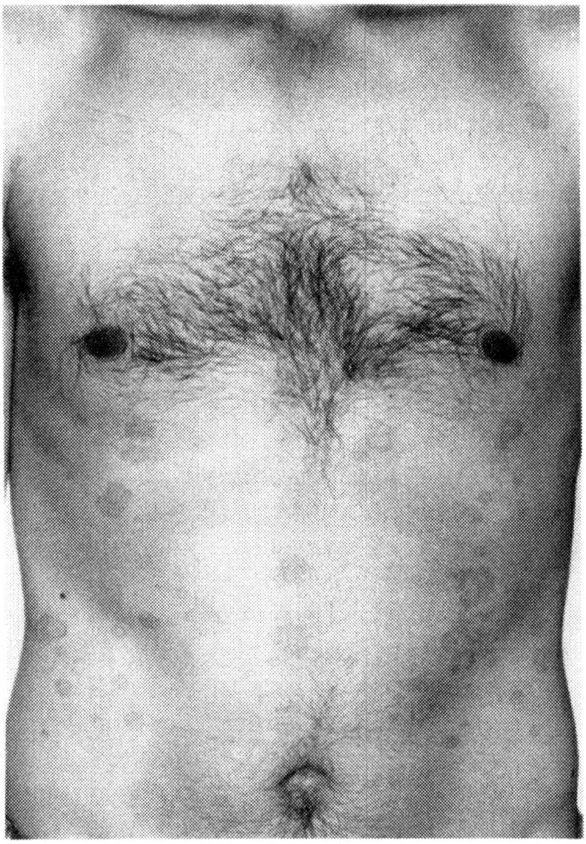

FIGURE 4.8 Pityriasis versicolor showing annular lesions on the patient's left

Pityriasis versicolor rarely itches and it is not often noticed by patients on systemic steroids. The author has a suspicion that skin changes in these patients are even quieter and less scaly than those in normal subjects but there is no histological or other evidence to support this suggestion. In patients whose infection has inadvertently been treated with topical steroids the rash may be very indistinct – a mixture of erythema, telangiectasia and fine scaling with poorly demarcated outlines difficult to diagnose or show photographically. A more distinct picture of atrophy almost exactly coextensive with the original eruption has also followed steroid applications in a few cases[16].

Prevalence and age variations

Since Marples' work in Samoa a prevalence of 30% or more has been accepted as realistic in tropical climates[2]. In temperate countries a level of infection of 1% or under has been suggested by clinic attendances but until recently no reliable figure was available. Now a large population survey of 28 000 subjects between the ages of 1 and 74 examined for skin diseases in different parts of the USA between 1971 and 1974 is available and shows a total prevalence of 0.84% for pityriasis versicolor, 1.09% in males and 0.6% in females[17]. The peak age group was 18–24 (prevalence of 1.78%) in this survey, which confirms the reports from clinical studies where the peak is usually put between 20 and 30[2,4,5] regardless of climate. As one might expect from the age studies of the normal flora pityriasis versicolor is rare in infancy though it does occur[18]. One suspects that maternal endocrine influences allow enough sebaceous activity to support a fairly heavy population of pityrosporum yeasts to persist for a while. In childhood, before puberty, the condition seems rarer still even among steroid-treated children. General experience also suggests that pityriasis versicolor is not a disease of the elderly in colder climates. In tropical countries the situation is very different; for example in Manono in Western Samoa 10% of children under 10, and 15% of adults over 60 had pityriasis versicolor when Marples and Bacon undertook their survey described by Marples[2] in the 1950s. Apart from two small Polish surveys there are no reports from temperate climes of comparable levels of infection at the extremes of life.

HOST/PARASITE CHANGES

The essential change which must occur for a person with normal skin to develop pityriasis versicolor is an increase in the number of yeast cells and the formation of large amounts of fungal mycelium. In their investigations of otherwise fit pityriasis versicolor patients McGinley *et al.* quantified the yeast cells (counting yeast with an attached bud as one) and filaments (single short units of mycelium)[8]. Having collected their samples using Williamson's open cup method, they counted fungal elements directly using a haemocytometer counting chamber – because cultural methods commonly used for bacterial work were unsatisfactory. In the affected skin their geometric mean counts (30 samples) were 156 000 yeast cells/cm^2 compared with 19 000/cm^2 for unaffected sites on pityriasis versicolor patients (23 samples) and 10 000/cm^2 in normal subjects. Their filament counts were 250 000/cm^2 in the lesions and 5800/cm^2 on unaffected sites. They quote no filament figures for normal subjects – most sampled showed none at all. The author observed a few filaments in stained skin scrapings in 8% of 50 subjects, 2–3 sites per subject[10], and on the unaffected skin of the trunk of 5 out of 12 pityriasis versicolor patients (2 samples per subject) or 9/12 patients if filaments seen in scalp samples are included[5]. Burke in her experimental pityriasis versicolor work noted filament formation prior to the development of clinical lesions, an observation familiar to the author in immunosuppressed patients with spontaneous infections[4].

In 1977 two independent groups, Dorn and Roehnert in Munich[19] and Nazzaro-Porro and her associates in Rome[20], published convincing evidence of yeast mycelial transformation in *Pityrosporum* species *in vitro* after years of failure by many other investigators. The German workers restricted their claims to *P. orbiculare* while the Italians refer to '*Pityrosporum* species'. Prior to 1977 most investigators who had isolated *P. orbiculare* from pityriasis versicolor had observed germ tube formation in many cultures and remarked that one or two isolates had been capable of producing a short length of mycelium at least on occasions. The author found that in *P. orbiculare* isolates from normal subjects those coming from sites showing filament formation on direct microscopy were nine times more likely to develop germ tubes than those collected from sites showing only yeasts[5]. This could of course be a carry-over effect reflecting differences in the earlier habitat (different skin conditions) rather than inherent differences in the yeast itself. But it does seem likely that some strains of *P. orbiculare* are particularly given to mycelial transformation and are therefore perhaps more pathogenic.

In any host/parasite relationship the influence of the other members of the microbial ecosystem should not be ignored. The interactions between *P. orbiculare* and the normal flora have been studied. The Philadelphia group demonstrated a statistically significant three-fold increase of aerobic bacteria (433/cm^2 rising to 1380/cm^2)[8]. Their figures for *P. acnes,* 11 900 rising to 39 900/cm^2, were of a similar order but because of wide individual variations between subjects were not statistically significant. They showed that these changes were of the same order as, and therefore likely to be due to, the accumulation of corneocytes which they concluded provided suitable conditions for the passive proliferation of these organisms.

More than 20 years ago Marples speculated on the relationship between *P. orbiculare* and *P. ovale* and considered whether the oval yeast might inhibit the round one – perhaps by direct competition for common nutrients[2]. In an attempt to pursue this point I attempted semi-quantitative estimates of oval and round yeast forms in skin scrapings from normal subjects with minor unrelated skin pathology and in the unaffected skin of pityriasis versicolor patients. On the trunk *P. ovale* was seen on 38/50 normals (76%) but only 3/12 (25%) of pityriasis versicolor patients. Among the normals 7/50 (14%) had an apparent predominance of oval yeasts on the trunk, whereas none of the three pityriasis versicolor patients out of 12 had more than a minority of oval yeasts. In scaly samples the pityriasis versicolor patients only differed in having a greater tendency to show some *P. orbiculare*, 12/12, compared to 68/100 for normals. These findings cannot be regarded as proof of any definite reciprocal relationship. They are of course compatible with the obvious proposition that if *P. orbiculare* is the usual cause of pityriasis versicolor a subject who already carries a normal flora in which that organism predominates is more likely to be afflicted than one whose predominant yeast is *P. ovale*.

If, however, one accepts the recent suggestions that *P. orbiculare* and *P. ovale* are simply different stages in the life cycle of the same organism, these observations require a somewhat different interpretation. Moreover there are, as already stated, some patients with pityriasis versicolor whose scrapings show filaments accompanied by clusters of oval yeasts. This finding may be more common among immunosuppressed patients but it certainly occurs in otherwise fit subjects and in both instances seems to be associated with a predominant oval yeast flora. Whether systemic steroids affect the ratio of oval to round yeasts, be this a question of the balance between two species or between two phases of a life cycle, is not presently clear. Nor is there any understanding of the way steroids might affect filament formation, though this does not depend on total numbers as filaments can be found before there has been any dramatic increase in the yeast population.

THE HOST RESPONSE

Histology and ultrastructure

In recent years transmission and scanning electron microscopy have supplemented the histological information previously established by conventional light microscopy. The main changes are a mild degree of acanthosis of the epidermis with hyperkeratosis. In some specimens the malpighian layer seems slightly spongiotic but often no such changes are seen. Similarly a very quiet inflammatory cell infiltrate in the lower dermis has been reported but it is often inapparent or absent. A lymphocytic and histiocytic perivascular infiltrate occurs in the upper dermis but is usually slight – as is any increased folding of the dermo-epidermal border. The site of fungal elements is normally the outer two-thirds of the stratum corneum, though some hyphae do penetrate deeper. Filaments can be seen to penetrate the keratinocytes though they may also insinuate themselves between cells. There is a clear zone around these mycelial elements implying the presence of a keratinase. On the whole the yeast cells are observed more superficially than the filaments and in particular they cluster in the infundibulum of the hair follicle. Presumably as a direct result of fungal damage the cells of the outer two-thirds of the stratum corneum are disrupted – some bent and upturned, others apparently exploded[11]. In some reports mention is made of clusters of Langerhans cells, but most authors concentrate on melanocyte and melanosome changes. In hyperpigmented lesions irregular pigment distribution is reported and in some cases the melanosomes are larger than normal. One recent publication points out that the colour change in hyperpigmented pityriasis versicolor cannot be entirely due to melanin, as a patient who developed it in skin devoid of melanocytes from longstanding vitiligo still had lesions darker than the surrounding skin[21]. While in general the living keratinocytes in the deeper parts of the epidermis appear to be relatively unaffected some reports do mention dark, apparently condensed, cells – thought to be altered keratinocytes[22]. In hypopigmented pityriasis versicolor there is a diminution in the number and size of melanosomes and maldistribution, as well as evidence of damage to melanocytes. If the mechanism of hyperpigmentation remains obscure, the changes seen in hypopigmentation lesions can now be better explained. Azelaic acid, produced by pityrosporum yeasts from fatty acids in sebum, is a dicarboxylic acid which is capable of inhibiting tyrosinase and it can also cause structural damage to hyperactive normal melanocytes. It has been used in the treatment of melanoma and can produce at least partial inhibition of melanoma cells[23].

Usually conventional histology shows no obvious parakeratosis and in one investigation, turnover studies using the Baker/Kligman technique seemed to be normal[22]. However, not only were corneocyte counts high, as would be expected, but the percentage of parakeratotic cells was much increased in the study of McGinley and his coworkers[8]. In pityriasis versicolor lesions the geometric mean count was 377 000 corneocytes/cm^2 compared to 124 000/cm^2 on unaffected skin, a three-fold increase, while the parakeratotic cells were eight times as numerous; 47 500/cm^2 compared to 5500/cm^2.

Fluorescence

The well-known yellow fluorescence under ultraviolet light, so useful clinically, revealing as it does many otherwise inconspicuous lesions, has hardly been investigated. The chemical composition of the substances concerned is not known, nor is its biological function. Fluorescence is not found when *P. orbiculare* or *P. ovale* grow *in vitro* but investigations with the scanning electron microscope and cathodoluminescence clearly demonstrated that the fluorescence resides in the fungal elements, both yeasts and filaments, and not in the host tissues[24]. It can be found in other scaly lesions which are heavily colonized by round or oval *Pityrosporum* species but which show no filament formation, for example in the greasy scales of non-inflammatory cradle cap, scalp scaling in some immunosuppressed patients and in some cases of petaloid seborrhoeic dermatitis. The follicles of the upper trunk may also fluoresce bright yellow in normal subjects, perhaps due to pityrospora in the infundibular region. Certainly this is a striking finding in certain patients with pityriasis versicolor after their lesions have been cleared with treatment. The author suspects, but cannot prove, that a high degree of yellow follicular fluorescence after treatment indicates a greater likelihood of relapse.

Immune responses

For over 20 years since the early work with fluorescent antibodies, it has been known that *Pityrosporum* species were capable of eliciting an immune response in man. It is now clear that *P. orbiculare*, though apparently a much weaker antigen weight for weight than, for example *Candida albicans,* is able to stimulate antibody production in experimental animals such as rabbits and guinea pigs. Healthy human subjects regularly show humoral and cellular responses to *Pityrosporum* species and pityriasis versicolor patients often have higher antibody levels than normal[25]. The antigenic structure of *P. orbiculare* is similar to *P. ovale*[26]. They undoubtedly have some antigens in common, and such differences as have so far been revealed are minor ones. *P. orbiculare* in common with various other micro-organisms is capable of activating complement by the alternative pathway but the significance of this finding remains unclear[27].

Although humoral antibody levels against *P. orbiculare* or *P. ovale* are often higher in patients with pityriasis versicolor, T-cell function as expressed by lymphocyte migration tests using *P. orbiculare* antigens may be reduced in many patients. In one study by Sohnle's group in Wisconsin there was a diminished production of leukocyte migration inhibition factor in response to *P. orbiculare* antigens but also to *Candida* antigens – an unexplained finding in the absence of any evidence of other infections or generally impaired immunity[28]. Recent work by the same team suggests that the impaired lymphocytic reactions are in part due to a diminished number of cells immunoresponsive to *P. orbiculare* compared to normal[29]. They discovered no evidence of active suppression of immune reaction as has been demonstrated in some other infections. These observations and the well-known lack of a significant degree of inflammation clinically and histologically in pityriasis versicolor should not be taken to mean there are no cellular responses. Lymphocytes are found, albeit in small numbers, in the epidermis and the upper dermis (perivascularly) and have been shown to be T-helper/inducer cells in the main, with small numbers of suppressor/cytotoxic cells[30]. It was also noted that there was a close spacial association between helper/inducer cells and HLA-DR reacting dendritic cells (Langerhans cells) in both the epidermis and dermis, though these authors were careful not to speculate as to the interpretation of this finding. Positive patch tests to *P. ovale* antigens in 10/10 psoriasis patients but only 2/10 controls have been reported[31]. The relevance of this intriguing finding remains obscure to the author but the fact that two positives among controls both turned out to have, or have had, seborrhoeic dermatitis is intriguing. Patch tests with *P. orbiculare* antigen appear not to have been done but might be predicted to be negative.

Susceptibility

None of the immunological studies so far performed has demonstrated a convincing explanation for the susceptibility some patients show towards *P. orbiculare* and their subsequent development of pityriasis versicolor. Clinical investigations have pointed to age, the peak being in the twenties in both temperate and tropical climates, and of course climate, season and even local environment as relevant factors[5]. Familial cases have been noted more commonly than conjugal ones – though the latter do occur and a genetic susceptibility factor remains possible but is not proven[4,5,7]. Even an established familial predisposition might be related to particular characteristics of the yeast rather than the host. Members of a family are likely to be colonized by the same strain of *P. orbiculare* early in life, the numbers increasing at puberty. Marriage would be less likely to permit the colonization of a new strain on an adult skin already possessing a fully established flora. Although the important US population survey[17] suggests a male predominance of nearly 2:1 this is not the case in all reports and is nowhere near the 9:1 male predominance of tinea cruris for example. The explanation remains unclear, and studies of yeast population densities to compare the sexes have not been published.

Among women pregnancy has been reported as a predisposing

factor but without convincing evidence[4]. Anecdotally oral contraception has been implicated[5]. The author, searching for evidence of filament formation microscopically in skin scrapings from women on the pill compared with age-matched controls, noted an apparent higher level, approximately 24% compared to 12% in the early part of the study. However, the difference became much less as the work progressed. This trend coincided with the move towards a lower oestrogen content in oral contraceptives. It is possible that higher oestrogen levels do influence the body's response to *P. orbiculare* but it would be wrong to draw any definite conclusions from this unpublished investigation.

Malnutrition is also commonly quoted as a predisposing factor but apart from its effects on immune function there is no obvious explanation and some of the reports on which this suggestion has been based failed to exclude climatic factors.

Prolonged treatment with systemic steroids was shown by Boardman and Malkinson in the 1960s to be associated with a 10% level of pityriasis versicolor in patients with inflammatory bowel disease[32]. Their control group of patients not so treated had no pityriasis versicolor. Burke noted the fluctuation of severity of the infection with fluctuating steroid dosage and achieved an experimental infection in a patient on systemic steroids having failed to do so in an untreated subject[4]. Subsequent surveys of steroid and azathioprine-treated patients – usually without control populations – have shown higher levels of pityriasis versicolor than expected. Repeated observations of patients in Cambridge have shown 74% of over 100 renal transplant recipients treated with prednisolone and azathioprine had pityriasis versicolor some time in the first 2 years of treatment. Most surveys based on a single observation have given lower figures similar to Boardman and Malkinson's 10%.

Other immunosuppressive regimens have not been studied so thoroughly. So far in Cambridge the only patient treated with cyclosporin who has developed obvious pityriasis versicolor also had systemic steroids not long before. Among the other immune disorders Sohnle quotes two patients with chronic mucocutaneous candidosis who also had pityriasis versicolor. Generally the literature on this condition does not reveal a particular predisposition to lipophilic yeast infections, perhaps because systemic amphotericin is used so frequently and suppresses the normal lipophilic yeast flora of the skin, thus serving as a prophylactic.

Quite recently Faergemann reported a significant association between seborrhoeic dermatitis and pityriasis versicolor[7]. Though never previously recognized and unconfirmed, his figures look convincing and are of interest in the light of recent work suggesting a causal role for *P. ovale* in seborrhoeic dermatitis. Prior to Faergemann's observations it was reasonable to assume that patients with pityriasis versicolor were subjects whose skin supported (for ill-understood reasons) a high *Pityrosporum* (*orbiculare*) flora which underwent yeast mycelial shift invoking a hyperkeratotic epidermal response with little inflammation. Conversely in some forms of seborrhoeic dermatitis the patient, perhaps supporting larger than average numbers of *Pityrosporum* species (*P. ovale*) could be thought of as having developed an inflammatory response to those organisms. If the round and oval forms are, as they seem to be, antigenically very similar one might expect the two conditions to be incompatible or if they did coexist the pityriasis versicolor lesions ought to be very inflamed. Faergemann's publication runs counter to this. Furthermore if the two forms of yeast are different stages in the life cycle of the same organism the situation becomes even more confusing. Seborrhoeic dermatitis and the role of *Pityrosporum* species and its aetiology is dealt with elsewhere (Chapter 5) but its common occurrence in AIDS, in which pityriasis versicolor seems not to be especially prevalent, and its relative absence in transplant patients, among whom pityriasis versicolor is common, raises new questions.

For nearly 25 years investigators have been attempting to explain the susceptibility to pityriasis versicolor in otherwise fit subjects in terms of fundamental differences in skin type biochemistry. Naturally lipid levels attracted early attention but neither Burke[4] nor Boardman and Malkinson[32] could find any difference between the surface lipids of normal subjects and those of the unaffected skin of patients with pityriasis versicolor. Burke did find some evidence that the yeasts and filaments within the lesions consumed lipid, but that was all. Her further studies on the amino acid concentrations of the epidermis revealed various differences between controls and unaffected skin of patients[33]. She felt the most significant findings were increased amounts of arginine and phenylalanine. Increased asparagine, glutamic acid and threonine with diminished isoleucine and leucine were, she believed, secondary effects resulting from heavy colonization of the skin. Others have reported that a total increase in epidermal amino acid levels is an important factor, and that alkali neutralization time and reduced water spreading are also significant, but these suggestions remain unconfirmed[6].

Coninx (personal communication, 1969) studied pityriasis versicolor patients, including some with steroid-induced infection, in an attempt to discover whether glucose metabolism is important in susceptibility. He could find no evidence to suggest that it was, and contrary to occasional anecdotal claims that diabetes predisposes to this infection none of his pityriasis versicolor patients had diabetes or increased glucose levels. Moreover other subjects with steroid-induced glycosuria did not have pityriasis versicolor.

Catterall *et al.* investigated lipase activity in vitro and *in vivo* in *P. orbiculare*[34]. They found evidence of enzymic activity in cultures but not in skin samples from pityriasis versicolor. They concluded therefore that the surface lipids were irrelevant in pityriasis versicolor. From earlier work quoted above we know that individual differences in lipid levels seem not to be important in susceptibility, but the tendency for pityriasis versicolor to attack lipid-rich areas, its age peak, and its almost complete absence from palms and soles, suggests that a relatively high baseline lipid requirement is necessary, if only to support an initial population of *P. orbiculare*.

The importance of extrinsic lipid applied to the skin, for example in the form of sun-tan oil, is often accepted as an important factor. The practice of oiling the skin is common in some tropical climates and may partially explain the higher levels of infection, but this is unproven. A recent publication showing that a common sun-tan oil contains several ingredients which will support the growth of *P. orbiculare in vitro* is of interest but its significance in practice is still uncertain[35]. This author found no evidence that yeasts ever contaminated the oil, and the possibility that an infection might be transmitted by such a vehicle seems remote.

To date the substances used by Dorn and Roehnert (glycine)[19] and by Nazzaro-Porro's team (cholesterol and cholesterol esters)[20] to induce yeast/mycelial transformation have not been shown to be important *in vivo*, though the latter are of course present on the skin.

In experimental infections with *P. orbiculare* occlusion has been found to be important[18]. Reports of localized pityriasis versicolor under situations where occlusion has occurred naturally also suggest that the accumulation of moisture and perhaps high CO_2 tensions may favour the development of pityriasis versicolor. Occlusion is of course helpful in experimental bacterial infections as well as those with *Candida* and dermatophytes. Whether excess sweating is the main explanation for the prevalence of pityriasis versicolor in the tropics remains unknown. Infection of flexural sites did seem to be particularly associated with exposure to hot environmental conditions in Cambridge patients, but this feature has not been emphasized by other writers[5]. The dry skin in renal failure (in which there is atrophy of sweat glands) supports a sparse flora of bacteria[6] and scrapings from such patients often show very few lipophilic yeasts. After successful renal transplantation the skin becomes more moist, presumably as the sweat glands recover. Certainly skin scrapings from such patients show numerous *Pityrosporum* yeasts regardless of whether pityriasis versicolor develops.

We still have much to learn about host susceptibility to this yeast but a skin with adequate lipid, perhaps with excess sweating, certainly a warm climate and immunosuppression with steroids all promote the yeast. Pityriasis versicolor may well be encouraged by exogenous lipid and possibly in women by pregnancy or high levels of oestrogen medication. Minor immune abnormalities may be significant and are perhaps genetically determined, but biochemical susceptibility factors have yet to be reliably defined.

TREATMENT

Pityriasis versicolor varies considerably in extent and chronicity. In some patients the rash will fluctuate with the season, clearing in winter. Spontaneous resolution with or without a seasonal bias is probably very common, particularly in temperate climates. Steroid-induced infections may fluctuate with dosage, as Burke[4] noted, and it is certainly not uncommon for pityriasis versicolor in transplant patients to clear as the steroid dose is reduced. Although many patients with mild infections are unaware of their condition and many steroid-treated patients preoccupied with their underlying disease, most of those who attend skin clinics complaining of the cosmetic disability are anxious for treatment and do not wish to wait for spontaneous regression.

Treatment is easier and relapses are less common in temperate climates than in tropical countries, but as in any infection in which the organism is a member of the normal skin flora, recurrence after apparent cure is likely to be a frequent problem regardless of the environmental conditions. Apart from withdrawal or reduction of immunosuppression there is as yet no effective way of increasing host resistance to *P. orbiculare*. Treatment therefore relies on suppression of the organism, usually with topical remedies. As the shampoo manufacturers have known for many years *P. ovale* is very sensitive *in vitro* to selenium sulphide and zinc pyrithione. The clinical evidence using anti-dandruff shampoos containing these substances suggests that *P. orbiculare* is susceptible to both compounds. It has long been known that pityrospora are naturally sensitive to the polyenes, nystatin and amphotericin, although a nystatin resistance may be induced experimentally in *P. ovale*. The broad-spectrum imidazole antifungals are also effective against *P. orbiculare*. It is doubtful if there is any clinically significant difference between clotrimazole, econazole, miconazole, sulconazole, tioconazole or ketoconazole (when used topically) – none has yet been shown. Few therapeutic trials have included lengthy follow-up data but in their powers to kill the yeasts there seems little to choose between the various remedies mentioned above. The main practical difference lies in the fact that selenium sulphide and zinc pyrithione are effective if used only once or twice (usually a week apart) and the polyenes and topical imidazoles seem to require twice-daily application for 2–3 weeks. This point has not been pursued in detail but most published trials have followed these schedules. The difference in cost between one or two applications of an anti-dandruff shampoo and twice-daily usage of antifungal creams for 2 weeks is considerable, particularly if the rash is extensive. Various other remedies have been shown to work, for example Whitfield's ointment, propylene glycol, and sodium thiosulphate.

Ketoconazole orally is effective within a few days, though more prolonged treatment probably gives better long-term results and intermittent treatment has been suggested to prevent relapse in tropical climates[36]. It should be said that the place of oral ketaconazole in the treatment of pityriasis versicolor is not yet clear. The rare but well-recognized side-effect of hepatic damage makes it seem an inappropriate choice in such a harmless condition. However, treatment schedules can be very short in this infection, and hepatic problems have nearly always occurred only after several weeks medication, so perhaps this may be an over-cautious attitude. Certainly for patients with very widespread rashes and with follicular lesions or marked hirsutism it has some attractive advantages. In tropical climates too its use is understandably more tempting.

In the author's view the application of selenium sulphide 2.5% in liquid shampoo form or zinc pyrithione at bedtime, washed off the next morning and repeated once, a week later, is the simplest and most effective remedy. A topical imidazole applied twice daily for 2 weeks is appropriate for tender sites such as the groin, face or pubic area. Occasionally in inflamed pityriasis versicolor, an uncommon condition, a few days' treatment with a cream containing an imidazole and 1% hydrocortisone night and morning will render the skin suitable for selenium or zinc pyrithione treatment. Without it soreness and irritation may result but these side-effects are very uncommon in ordinary patients in temperate climates.

With an effective treatment the fungal elements soon disappear and the scaling clears, but it may be many months, even years, before repigmentation returns in hypopigmented lesions. Attempts have been made to speed the process using PUVA but are of no proven value and are surely unjustified in most patients. Clinical experience in temperate zones strongly suggests that repigmentation, though slow, does always occur following successful initial therapy and prompt retreatment of relapses. Whether pityrospora can ever induce permanent depigmentation is not known with certainty and cannot therefore be totally discounted. In practice, however, a patient but confidently optimistic attitude seems justified. At least this is the case in cold climates and among compliant patients who are carefully followed up.

REFERENCES

1. Sloof, W. Ch. (1970). *Pityrosporum* Sabouraud. In Lodder, J. (ed.). *The Yeasts*. 2nd Edn, pp. 1167–1186. (Amsterdam: North Holland)
2. Marples, M. J. (1965). Fungi inhabiting the skin: *Pityrosporum* and other yeasts. In *The Ecology of Human Skin*, Ch. 23, pp. 407–427. (Springfield, Ill: Thomas)
3. Gordon, M. A. (1951). Lipophilic yeastlike organisms associated with tinea versicolor. *J. Invest. Dermatol.*, **17**, 267–272
4. Burke, R. C. (1961). Tinea versicolor susceptibility factors and experimental infection in human beings. *J. Invest. Dermatol.*, **36**, 389–402
5. Roberts, S. O. B. (1969). Pityriasis versicolor. A clinical and mycological investigation. *Br. J. Dermatol.*, **81**, 315–326
6. Noble, W. C. (1981). *The Microbiology of Human Skin*. 2nd Edn. (London: Lloyd-Luke)
7. Faergemann, J. and Fredriksson, T. (1982). Tinea versicolor. Some new aspects on etiology, pathogenesis and treatment. *Int. J. Dermatol.*, **21**, 8–11
8. McGinley, K. J., Lantis, L. R. and Marples, R. R. (1970). Microbiology of tinea versicolor. *Arch. Dermatol.*, **102**, 168–171
9. Marples, R. R., Downing, D. T. and Kligman, A. M. (1972). Influence of *Pityrosporum* species in the generation of free fatty acids in human surface lipids. *J. Invest. Dermatol.*, **58**, 155–158
10. Roberts, S. O. B. (1969) *Pityrosporum orbiculare*: incidence and distribution on clinically normal skin. *Br. J. Dermatol.*, **81**, 264–269
11. Pierard, J. and Dockx, P. (1972). The ultrastructure of tinea versicolor and *Malassezia furfur*. *Int. J. Dermatol.*, **11**, 116–124
12. Brotherton, J. (1967). Lack of swelling and shrinking of *Pityrosporum ovale* in media of different osmotic pressures and its relationship with survival in the relatively dry conditions of the scalp. *J. Gen. Microbiol.*, **48**, 305–308
13. Ranjandiche, M. (1976). Polymorphisme de *Pityrosporum ovale* (Bizzozero) Castellani & Chalmers *in vivo* et *in vitro*. *Bulletin Societé Mycol. Med.*, **V**, 79–84
14. Salkin, I. F. and Gordon, M. A. (1977). Polymorphism of *Malassezia furfur*. *Can. J. Microbiol.*, **23**, 471–475
15. Vanbreuseghem, R. and de Tiege, R. (1952). Contribution à l'étude du pityriasis versicolor et de *Pityrosporum ovale*. *Ann. Soc. Belg. Med. Trop.*, **32**, 521–528
16. Tatnall, F. M. and Rycroft, R. J. G. (1985). Pityriasis versicolor with cutaneous atrophy induced by topical steroids. *Clin. Exp. Dermatol.*, **10**, 258–261
17. National Health Survey (1978). Skin conditions and related need for medical care among persons 1–74 years. United States 1971–1974. U.S. Dept. of Health Education and Welfare DHEW Pub. No. (PHS) 79–1660. Office of Ass. Sec. for Health, National Center for Health Statistics, Hyattsville Md.
18. Faergemann, J. (1979) Tinea versicolor and *Pityrosporum orbiculare*. Mycological investigations, experimental infections and epidemiological surveys. *Acta Dermatol. Venereol. (Stockh.) Suppl.*, **86**
19. Dorn, M. and Roehnert, K. (1977). Dimorphism of *Pityrosporum orbiculare* in a defined culture medium. *J. Invest. Dermatol.*, **69**, 244–248
20. Nazzaro-Porro, M., Passi, S., Caprilli, F. and Mercantini, R. (1977). Induction of hyphae in cultures of *Pityrosporum* by cholesterol and cholesterolesters. *J. Invest. Dermatol.*, **69**, 531–534
21. Dotz, W. I., Hendrikson, D. M., Yu, G. S. M. and Galey, C. I. (1985). Tinea

versicolor. A light and electron microscopic study of hyperpigmented skin. *J. Am. Acad. Dermatol.*, **12,** 37–44

22. Charles, C. R., Sire, D. J., Johnson, B. J. and Beidler, J. G. (1973). Hypopigmentation in tinea versicolor: a histochemical and electron microscopic study. *Int. J. Dermatol.*, **12,** 48–58
23. Robins, E. J., Breathnach, A. S., Bennett, D., Ward, B. J., Bhasin, Y., Ethridge, L., Nazzaro-Poro, M., Passi, S. and Picardo, M. (1985). Ultrastructural observations on the effect of azaleic acid on normal human melanocytes and a human melanoma cell line in tissue culture. *Br. J. Dermatol.*, **113,** 687–697
24. Carteaud, J. P., Muir, M. D. K. and Grant, P. R. (1972) Pityriasis versicolor. Emissive and cathodoluminescence examination in the scanning electron microscope. *Sabouraudia,* **10,** 143–146
25. Da Mert, G. J., Kirkpatrick, C. H. and Sohnle, P. G. (1980) Comparison of antibody responses in chronic mucocutaneous candidiasis and tinea versicolor. *Int. Arch. Allergy Appl. Immunol.*, **63,** 97–104
26. Tanaka, M. and Imamura, S. (1979). Immunological studies on *Pityrosporum* genus and *Malassezia furfur. J. Invest. Dermatol.*, **73,** 321–324
27. Sohnle, P. G. and Collins-Lech, C. (1983). Activation of complement by *Pityrosporum orbiculare. J. Invest. Dermatol.*, **80,** 93–97
28. Sohnle, P. G. and Collins-Lech, L. (1978). Cell-mediated immunity to *Pityrosporum orbiculare* in tinea versicolor. *J. Clin. Invest.*, **62,** 45–53
29. Sohnle, P. G. and Collins-Lech, C. (1982). Analysis of the lymphocyte transformation response to *Pityrosporum orbiculare* in patients with tinea versicolor. *Clin. Exp. Immunol.*, **49,** 559–564
30. Scheynius, A., Faergemann, J., Forsum, U. and Sjöberg, O. (1984). Phenotypic characterisation *in situ* of inflammatory cells in pityriasis (tinea) versicolor. *Acta Dermatol. Venereol.*, **64,** 473–479
31. Lober, C. W., Belew, P. W., Rosenberg, E. W. and Bale, G. (1982). Patch tests with killed sonicated microflora in patients with psoriasis. *Arch. Dermatol.*, **118,** 322–325
32. Boardman, C. R. and Malkinson, F. D. (1962) Tinea versicolor in steroid treated patients. *Arch. Dermatol.*, **85,** 44–52
33. Burke, R. C. (1962). Investigations in tinea versicolor: lipid and amino acid studies. *Yale J. Biol. Med.*, **35,** 206–221
34. Catterall, M. D., Ward, M. E. and Jacobs, P. (1978). A reappraisal of the role of *Pityrosporum orbiculare* in pityriasis versicolor and the significance of extracellular lipase. *J. Invest. Dermatol.*, **71,** 398–401
35. Roed-Petersen, Jytte (1980). Tinea versicolor and body lotions. *Acta Dermatol. Venereol. (Stockh.)*, **66,** 439–440
36. Hay, R. J. and Midgley, G. (1984) Short course ketoconazole therapy in Pityriasis versicolor. *Clin. Exp. Dermatol.*, **9,** 571–573

5

SEBORRHOEIC DERMATITIS

F. A. IVE and C. M. E. ROWLAND PAYNE

SEBORRHOEIC DERMATITIS IS A FUNGAL DISEASE

The fact that seborrhoeic dermatitis is a fungal disease will come as a considerable surprise to many modern dermatologists. It has long been known that large quantities of spores of *Pityrosporum* species overgrow in involved skin. Since these are recognized as saprophytic organisms found in lesser numbers in all normal individuals, consensus had it that their profusion resulted from the veritable beanfeast of sebum, sweat and squames provided by the 'seborrhoeic' environment.

Older and wiser continental dermatologists can now be vindicated for their belief that seborrhoeic dermatitis is for the most part infective.

Rivolta[1] in 1873 first visualized the organism from the crusted skin that flaked within his venerable beard. He called it *Cryptococcus psoriasis* and regarded it as the cause of that disease. He would have been interested in the recent work of Rosenberg[2] and others[3,4] demonstrating overgrowth of *Pityrosporum* yeasts in psoriasis and considerable improvement after their removal.

Malassez[5] considered the organism to be responsible for scalp scaling, and in 1904 Sabouraud[6] declared his belief that *Malassezia ovalis*, later to be known as *Pityrosporum ovale*, was the yeast responsible for seborrhoeic dermatitis. From that point on it was down the hill all the way for the infective theory. Proponents of the purely saprophytic role argued that the organism evoked no host response in normal subjects whilst, of course, the well-recognized fact that it can be found in its spore form in all humans means that Koch's postulates could never be fulfilled.

In the past few years it has been possible to investigate the disease using the new drug ketoconazole. This drug is a broad-spectrum antifungal with little antibacterial activity and no cytostatic effect. If it could 'cure' seborrhoeic dermatitis then maybe Koch would not expostulate too forcefully.

The generic term seborrhoeic dermatitis was first used by Unna in 1887[7] to describe a variety of clinical entities which have since been re-fragmented into such diagnostic terms as seborrhoea, dandruff, scurf and pityriasis, with a sufficiency of suffixes such as sicca, corporis, capitis, oleosa steatoides and circinata which have proved helpful only in adding confusion to the issue. This fragmentation was almost inevitable because of the absence of either a distinctive diagnostic test for seborrhoeic dermatitis, or a specific experimental animal model. It is only by applying the results of recent studies with antifungal preparations that we have been able to make clinically unifying conclusions.

Because of doubts as to its aetiology seborrhoeic dermatitis has been allowed to languish as an unprecise *entrepôt* of a diagnosis where each individual practitioner has his own conception of its non-entity and some, discreetly, even question its very existence.[8]

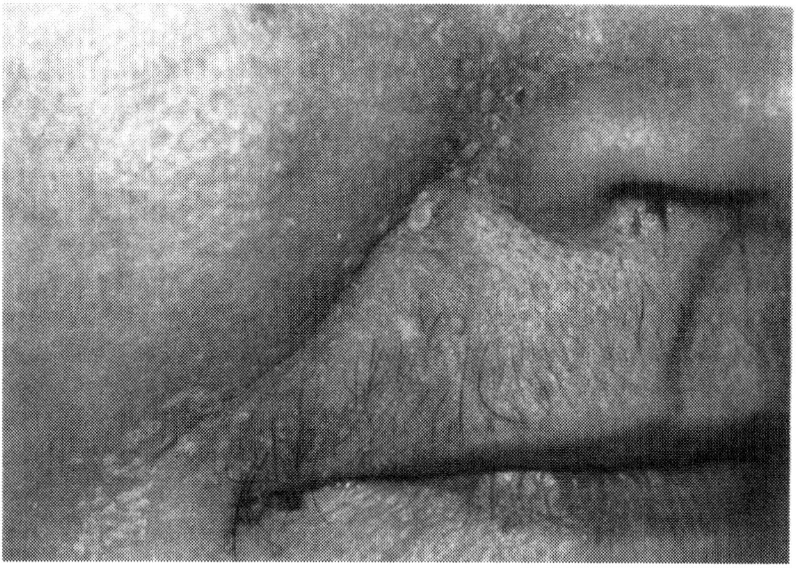

FIGURE 5.1 Seborrhoeic dermatitis involving paranasal gutter and nasolabial fold

Investigation with ketoconazole has demanded that only watertight universally accepted examples of seborrhoeic dermatitis have been investigated. These variants of the disorder have been extracted from the *entrepôt* and will be carefully described as part of the total entity. Other presumed variants must be left in the *pôt* for the time being, but are worthy of the same type of investigation, although recent doubts in the safety of systemic ketoconazole are likely to be a limiting factor.

In an attempt to define uniform criteria for the diagnosis of seborrhoeic dermatitis, the condition has been divided into clinical seborrhoeic dermatitis (Table 5.1), complications, and a number of variants. Clinical seborrhoeic dermatitis and its variants respond to fungicides but the superimposed complications do not.

TABLE 1 Diagnostic features of classical seborrhoeic dermatitis

I	Erythematous squamous eruption, without scarring at *one* or more of the following six groups of sites:
	(a) scalp and/or hairline
	(b) eyebrows and/or glabella
	(c) paranasal gutters and/or nasolabial folds (Fig. 5.1)
	(d) cheeks and nasal bridge in a butterfly distribution
	(e) periauricular
	(f) presternal (often in a petaloid configuration)
II	Post-pubertal onset
III	Rapid response to topical (or systemic) pityrosporicidal therapy

CLASSICAL SEBORRHOEIC DERMATITIS

The scalp is the most commonly involved site. It is affected by a diffuse yellowish greasy scaling, associated with some itching. Margination of the eruption may occur with extension of it beyond the hairline, but mostly it remains discrete within the hair-bearing area; thus the bald tonsure of the ageing male is spared.

Much energy has been expended in the past in arguing for a distinction between dandruff and scalp seborrhoeic dermatitis. The argument revolves round the near universality of the dry loose flakes of dandruff compared with the rarer yellow crusting scale of seborrhoeic dermatitis.

The whole argument has become academic, sterile and of no clinical importance since everyone is agreed that *Pityrosporum* spores are present in large numbers in both conditions, and it has been shown that both conditions clear rapidly on treatment with either systemic or topical ketoconazole[9–11]. In practice dandruff and seborrhoeic dermatitis may be regarded as a continuum.

Spread beyond the confines of the scalp appears to follow a gravitational pattern, lending credence to the folklore which claims that dandruff is infectious. The eyebrows, nasolabial folds, cheeks or chest anteriorly, and the ears laterally, are usually sequentially involved before the body folds. The clinical appearance within the hair-bearing areas can be either of dry grey flaking or of greasy yellowish scales. In non-hair-bearing sites the appearance is most commonly that of intermittently itchy ill-defined macular lesions with dry scaling. Sometimes coalescing follicular papules with greasy peripheral scaling may form multiple circinate patches often described as petaloid in appearance by generations of imaginative dermatologists.

COMPLICATIONS OF SEBORRHOEIC DERMATITIS

Secondary changes in seborrhoeic dermatitis do not respond to fungicides, and their recognition is important if therapeutic disillusionment is to be avoided.

Secondary bacterial infection

Ingram[12] termed this complication ‘pyogen-prone’ and described the end result as a ‘sticky, oozing, pustular, scabbed and foul eczematous dermatitis ... particularly affecting the head and about the ears, the eyebrows, eyelids and beard area’. Clearly antibacterial medications are the first-line agents to be used in such a situation.

Flexural intertrigo

May involve both the great wet flexures (axillae, groins, intergluteal and inframammary) and the minor wet flexures (mouth angles, umbilical and toe webs). All these warm damp macerated zones are fertile soil for the growth of candida, dermatophytes and staphylococci. Treatment with drying agents such as the rosaniline dyes (gentian violent or brilliant green) is often helpful in these aspects of the disorder.

Lichenification (eczematization)

Constant rubbing of irritative lesions, particularly of the scalp and ears, can lead to thickened dry finely scaling plaques extending over the entire scalp. Many dermatologists mistakenly regard this pattern of scalp eczema as a form of classical seborrhoeic dermatitis. Ablation of the *Pityrosporum* spores in this situation produces disappointing results. Topical steroids, preferably in association with a keratolytic preparation such as 2% sulphur and salicylic acid cream, seems the treatment of choice.

Otitis externa

This represents a complicated mixture of seborrhoeic dermatitis associated with secondary bacterial and fungal invasion. Poor drainage and stagnation of secretions within the external canal compound the problems of secondary invasion. Eczematization and rubbing produce lichenification, which may in its turn be compounded by medicament sensitivity, lymphoedema, and deformity of cartilage. These further aggravate problems of drainage and secondary infection so that simple treatment of pityrosporon infection cannot be expected to provide benefit. A whole panoply of dermatological expertise may be required. Bacteriology, mycology, and patch testing are necessary investigations followed by appropriate systemic or topical antibiotics, antifungals, and steroid preparations. Once the acute attack is under control instillation of a drying agent such as 6% aluminium acetate eardrops is often sufficient to control relapses within the canal, whilst the application of steroid/antifungal creams such as hydrocortisone with clotrimazole or miconazole to the outer ear suppresses further itching and eczematization.

Associated psoriasis

It can be difficult to tell seborrhoeic dermatitis and psoriasis apart. Frequently they appear to coexist to their own mutual benefit. The psoriasis tends to 'Koebnerise' into areas of undoubted pre-existing seborrhoeic dermatitis. The preferred diagnosis in such situations has been designated as sebopsoriasis or even seborrhiasis. It is presumably in psoriatic cases of such a type that various authorities have had so much success with antifungal preparations[2–4]. We would certainly commend the addition of a topical antifungal in this variant of psoriasis of the face, scalp, or upper trunk.

FUNGICIDE-RESPONSIVE VARIANTS OF SEBORRHOEIC DERMATITIS

Variants of seborrhoeic dermatitis known to be responsive to antifungal agents:

Seborrhoeic blepharitis (Fig. 5.2)

This condition, characterized by yellow crusted scaling of the eyelid margins, is unusual in that it can affect the pre-pubertal child. Grease glands involved in tear lubrication function independently of sex hormone level and can support local fungal overgrowth in childhood. Management is usually very difficult and patients tend to present their encrusted sticky eye margins to the relatively ineffective ministrations both of dermatologists and ophthalmologists, to the mutual dissatisfaction of all concerned. Topical antifungal therapy is rapidly effective although treatment is hampered by the absence of any suitably sterilized ophthalmic preparation. Sterilized miconazole eye ointment is effective, but available only by special request from the manufacturers. Some enterprising company will, we trust, soon appreciate the existence of this profitable gap in the market.

Pityrosporum folliculitis[9,13] (Fig. 5.3)

This is a common low-grade folliculitis of the trunk which is frequently associated with seborrhoeic dermatitis. Though long recognized by

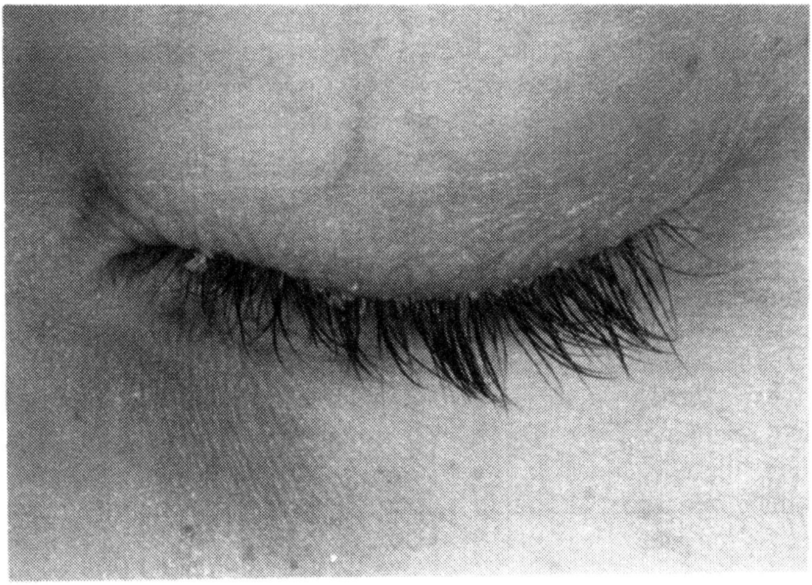

FIGURE 5.2 Seborrhoeic blepharitis

FIGURE 5.3 *Pityrosporum* folliculitis over trunk

the term seborrhoeic folliculitis it has recently been shown to be associated with the presence of *Pityrosporum* spores within the follicles and to respond to imidazole therapy.[9] *Pityrosporum* folliculitis is seen exclusively in pubertal and post-pubertal individuals of either sex. The primary lesion is a small pink follicular papule or white pustule. These are widely distributed across the upper trunk and shoulders, and may be associated with some slight scaling of the involved skin.

The disorder is frequently mistaken for mild acne, but is easily distinguished by the facial sparing, the absence of comedones, cysts or scars and failure to respond to antibiotics. The age group involved is usually postadolescent, and mild itching is common.

Treatment with systemic ketoconazole is effective, but cannot be safely recommended[14]. Topical ketoconazole as a 2% cream, or selenium sulphide shampoo as a cleansing agent, is usually adequate to produce clearance, although, as with seborrhoeic dermatitis, recurrence is the rule.

OTHER VARIANTS OF SEBORRHOEIC DERMATITIS

Infantile seborrhoeic dermatitis

Infantile seborrhoeic dermatitis presents very early in life, usually in the first month, and clears spontaneously by the third month; relapse is unusual. Its onset coincides exactly with the miniature puberty of the newborn when the influences of maternal and placental hormones are maximal.

This disorder fairly closely resembles post-pubertal seborrhoeic dermatitis although follicular lesions are rarely seen. Its relationship to adult disease is obscure and no satisfactory information on the *Pityrosporum* is available. *Candida albicans* has been grown from the skin of many of these infants[15] and it may be that this organism is acting as the infective agent. The disease is relatively non-itchy and the infant appears to be very little troubled by what can be an extensive and alarming eruption which commonly starts in the napkin area with lesions rapidly developing on the scalp, neck, and around the ears (Fig. 5.4); widespread greasy yellow scaling over the face and body may occur.

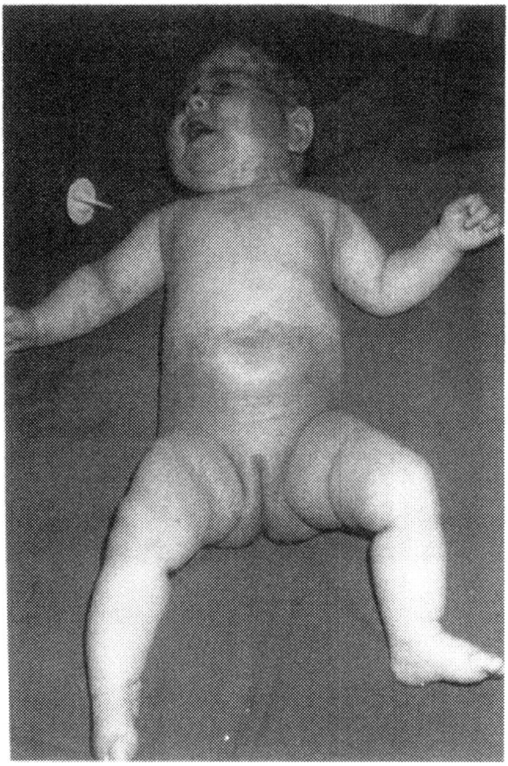

FIGURE 5.4 Infantile seborrhoeic dermatitis – typical widespread involvement

Secondary bacterial infection is the only important complication. Coloured children may develop marked post-inflammatory depigmentation which is temporary, and parents can be reassured (Fig. 5.5). When circinate plaques of a psoriasiform nature develop it has been called napkin psoriasis, which may represent sebopsoriasis of infancy (Fig. 5.6). Whether children with infantile seborrhoeic dermatitis develop classical seborrhoeic dermatitis after puberty, in the manner of infantile acne sufferers, is unknown. We do not share the opinion of Vickers that they are necessarily atopic[16].

Treatment can reduce the duration of the eruption, and weak topical steroids in association with nystatin and a topical antibacterial antibiotic are often helpful. No information is currently available on the absorption of imidazoles through infantile skin, so the use of topical

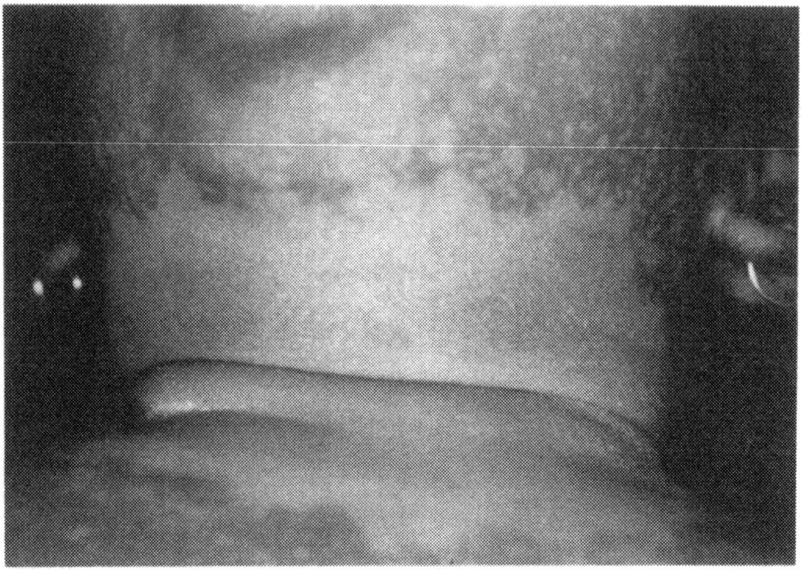

FIGURE 5.5 Infantile seborrhoeic dermatitis – post-inflammatory depigmentation over posterior scalp and neck

preparations of hydrocortisone in combination with miconazole or clotrimazole cannot yet be recommended.

Pityriasiform seborrhoeide

This is sometimes known as pityriasis corporis or pityriasiform eczematide. It is a prolonged and persistent eruption of adults which clinically resembles pityriasis rosea, although the herald patch is lacking and the distribution is more widespread, extending down the limbs and clustering around the body folds. Early lesions are often oval and erythematous with a marginated collarette of scale, older lesions may take on a slightly brownish hue and fine branny scales – which spread outwards into annular configurations.

Nothing is known of the nature of these lesions from the mycological standpoint, and further study is needed. They respond well to topical steroids mixed with chinoform but tend to relapse from time to time.

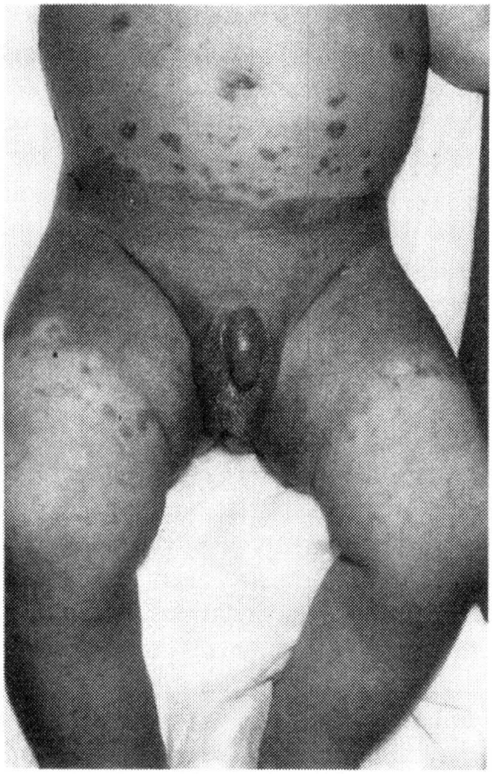

FIGURE 5.6 Napkin psoriasis – well-defined eruption with psoriasiform spread

DISCUSSION

The term seborrhoeic dermatitis is a misnomer, although there is no doubt that seborrhoeic dermatitis has a predilection for those areas in which the density of pilosebaceous glands is high. Burton and Pye[17] have shown that the quantity of sebum produced in affected sites and individuals may be small, and it must be accepted that seborrhoea in its literal sense is not a feature of the disease.

MICROBIOLOGY OF SEBORRHOEIC DERMATITIS

There is general agreement that the main organisms found in the normal scalp are aerobic bacilli, anaerobic bacilli (mainly *Pro-*

prionibacterium acnes), and *Pityrosporum ovale*. In the presence of seborrhoeic dermatitis the absolute number of organisms is increased with a disproportionate overgrowth of aerobes and fungi[18]. Direct microscopy reveals vast numbers of *Pityrosporum ovale*, mainly in its spore or yeast form, with very little evidence of the filamentous or mycelial form so commonly seen in pityriasis versicolor. We draw no distinction between the oval and round forms of spore, and would agree with Faergemann[19] that *Pityrosporum ovale* and *orbiculare* are mutually interchangeable. Apart from a few early and apparently successful implantation and colonization experiments[20,21] the investigation of the relationship between *Pityrosporum* and seborrhoeic dermatitis has been restricted largely to the effect of antifungal agents on the clinical condition and relating this to spore density. Nearly all studies, whatever the antifungal used, showed an improvement in rash associated with a decrease in spores. Nystatin[22], econazole[23] selenium sulphide[24] or zinc pyrithione[25] are all effective.

Early opponents of the fungal concept argued that since Koch's postulates were not fulfilled no relationship could be inferred[26]. In modern times the principal opponents have been Kligman and his colleagues[24]. They regard dandruff as a hyperproliferative condition of the scalp of unknown aetiology which is secondarily colonized by *Pityrosporum ovale* because of the favourable habitat. Clearance of symptoms and eruption with selenium sulphide shampoos was attributed not to the known antifungal qualities of this compound but to mysterious cytostatic qualities of dubious clinical relevance. Their arguments are poorly supported by evidence and have been shown by Shuster to 'verge on the absurd'[27]. The fact that their beliefs prevail in Anglo-American dermatology can only be a tribute to their mastery of the art of communication. The availability of ketoconazole has made it possible to answer some of the problems of seborrhoeic dermatitis and made many previous arguments redundant.

Ketoconazole (Nizoral) is an antifungal antibiotic designed for systemic administration. It has no measurable cytostatic effect but is active against a wide range of yeasts and fungi including keratinophilic ringworm organisms and non-keratinophilic *Candida* and *Pityrosporum* species. It was introduced into Britain in 1980 and, before it receded under its hepatotoxic cloud, it was used in the management and definition of firstly pityrosporum (seborrhoeic) folliculitis[14] and later in pure seborrhoeic (pityrosporum) dermatitis[9]. In both these double-blind controlled cross-over studies 200 mg per day of oral ketoconazole was shown to clear the eruption within 2–3 weeks. In all cases symptoms and signs recurred within 2 months.

Unfortunately ketoconazole has been implicated in the production of fatalities due to hepatic damage and should probably no longer be systemically administered except in life-threatening situations[28].

The drug is now available commercially as a 2% cream. Recent trials with the cream[10,11] and with an experimental shampoo formulation[10,29] have shown these to be equally effective in the treatment of what we have defined as classical seborrhoeic dermatitis. Just as with oral ketoconazole, symptoms and signs recurred within 2 weeks to 2 months in virtually all patients, implying the existence of a primary host defect enabling a common skin saprophyte to assume a parasitic role; or in alternative phraseology – testifying to the fertile soil of the seborrhoeic skin and the ubiquity of the *Pityrosporum* species.

Disturbance of a host–saprophyte relationship can occur in a variety of ways. It is unlikely that an immunological mechanism plays any major part in the disorder in that most sufferers are in good general health and free from the hallmarks of immunological impairment. Recent reports[30–32] on victims of acquired immune deficiency syndrome (AIDS) have indicated a high incidence of what appears to be seborrhoeic dermatitis from an early stage in their disease. No mycological studies were done, but seborrhoeic dermatitis is sometimes the earliest indicator of impaired immunity.

In the normal subject seborrhoeic dermatitis occurs only in post-pubertal adults (and possibly in the mini-puberty of infants) and never in eunuchs. This would imply that sex hormones have a specific effect on skin which is essential to the production of a welcoming milieu for pityro spores. Unna thought that an alteration in sebum could be responsible[7] but no substantiating quantitative[17] or qualitative[33] differences can be found. There is a general belief that sebum has antibacterial and antifungal properties[34,35] but it is difficult to find firm evidence supporting the claim. Kligman[36] has argued that the presence of corneocytes inhibits any possible antifungal effects of sebum. At present, while we must admit to complete ignorance as to the precise nature of the host defect in seborrhoeic dermatitis, the situation is no longer as depressing as a recent quotation might indicate[37]: 'The seborrhoeic state is a legacy which the patient can only accept. Its consequences can be reduced by regular supervision, sufficient sleep and relaxation and by a balanced diet.'

REFERENCES

1. Rivolta, S. (1873). *Parassiti Vegetali,* 1st Edn., pp. 469–471. (Torino, Italy: Giulio Speirani)
2. Rosenberg, E.W. and Belew, P.W. (1982). Improvement of psoriasis of the scalp with ketoconazole. *Arch. Dermatol.,* **118,** 370–371
3. Doring, H.F. (1984). Treatment of sebopsoriasis. *Dermatologica,* **169** (suppl. 1), 125–133
4. Farr, P.M., Krause, L.B., Marks, J.M. and Shuster, S. (1985). Response of scalp psoriasis to oral ketoconazole. *Lancet,* **2,** 921–922
5. Malassez, L. (1874). Notes sur le champignon de la pilade. *Arch. Physiol. Norm. Pathol.,* **1,** 203–212
6. Sabouraud, R. (1904). Pityriasis et alopecies peliculaire. *Les maladies desquamatives.* 1st Edn., p. 295. (Paris: Massonet)
7. Unna, P.G. (1887). Seborrhoeic eczema. *J. Cutan. Dis.,* **5,** 449–453
8. Steigleder, G.K. (1981). *Therapie der Hautkrankheiten.* 2nd Edn., pp. 92, 103. (Stuttgart: Thieme)
9. Ford, G.P., Farr, P.M., Ive, F.A. and Shuster, S. (1984). The response of seborrhoeic dermatitis to ketoconazole. *Br. J. Dermatol.,* **111,** 603–607
10. Green, C.A., Farr, P.M. and Shuster, S. (1986). Response of seborrhoeic dermatitis of the face and scalp to topical ketoconazole. *Br. J. Dermatol.* (awaiting publication)
11. Skinner, R.B., Noch, P.W., Taylor, R.M., Zanolli, M.D., West, S., Guin, J.D. and Rosenberg, E.W. (1956). Double blind treatment of seborrhoeic dermatitis with 2% ketoconazole cream. *J. Am. Acad. Dermatol.,* **12,** 852–856
12. Ingram, J.T. (1939). The seborrhoeic diathesis. *Br. Med. J.,* **2,** 5–8
13. Bäck, O., Faergemann, J. and Hornqvist, R. (1985). Pityrosporum folliculitis; A common disease of the young and middle aged. *J. Am. Acad. Dermatol.,* **12,** 56–61
14. Ford, G.P., Ive, F.A. and Midgley, G. (1982). Pityrosporum folliculitis and ketoconazole. *Br. J. Dermatol.,* **107,** 691–695
15. Beare, J.M., Cheeseman, E.A. and Mackenzie, D.W.R. (1968). The association between *Candida albicans* and lesions of seborrhoeic eczema. *Br. J. Dermatol.,* **80,** 675–681
16. Vickers, C.F.H. (1980). The natural history of atopic eczema. *Acta Dermatovener.* (*Stockh.*), Suppl. 92, 113
17. Burton, J.L. and Pye, R.J. (1983). Seborrhoea is not a feature of seborrhoeic dermatitis. *Br. Med. J.,* **286,** 1169–1170
18. Priestly, G.C. and Savin, J.A. (1976). The microbiology of dandruff. *Br. J. Dermatol.,* **94,** 469–471
19. Faergemann, J., Tjernlund, U., Scheynius, A. and Bernander, S. (1982). Antigenic similarities and differences in genus pityrosporum. *J. Invest. Dermatol.,* **78,** 28–31

20. Macleod, J. M. H. and Dowling, G. B. (1982). An experimental study of the pityrosporum of Malassez: its morphology, cultivation and pathogenicity. *Br. J. Dermatol. Syph.*, **11**, 139–148
21. Moore, M., Kile, R. L., Engman, M. F. Jr and Engman, M. F. (1936). *Pityrosporum ovale*. Cultivation and possible role in seborrhoeic dermatitis. *Arch. Derm. Syph.*, **33**, 457–472
20. Vanderwyk, R. W. and Roia, F. C. (1964). The relationship between dandruff and the microbial flora of the human scalp. *J. Soc. Cosmetic Chem.*, **15**, 761–768
23. Aron-Brunetiere, R., Dompmartin-Pernot, D. and Droubet, E. (1977). Treatment of pityriasis capitis (dandruff) with econazole nitrate. *Acta Dermatovenereol.* (Stockholm), **57**, 77–80
24. Leyden, J. J., McGinley, J. K. and Kligman, A. M. (1976). Role of micro-organisms in dandruff. *Arch. Dermatol.*, **112**, 333–338
25. Imokawa, G., Shimizu, H. and Okamoto, K. (1981). Antidandruff mechanism of zinc pyrithione. Proceedings of the Meeting of the Society of Cosmetic Chemists Washington, DC
26. Martin-Scott, I. (1952). The *Pityrosporum ovale*. *Br. J. Dermatol.*, **64**, 257–273
27. Shuster, S. (1984). The aetiology of dandruff and the mode of action of therapeutic agents. *Br. J. Dermatol.*, **111**, 235–242
28. Goldberg, A. (1985). *Oral ketoconazole and liver damage*. Committee Safety of Medicines. Adverse Reaction Series. No. 19
29. Carr, M. M., Pryce, D. and Ive, F. A. (1986). Response of seborrhoeic dermatitis of the scalp to topical ketoconazole. *Br. J. Dermatol.* (awaiting publication)
30. Farthing, C. F., Staughton, R. C. D. and Rowland Payne, C. M. E. (1985). Skin disease in homosexual patients with acquired immune deficiency syndrome (AIDS) and lesser forms of human T-cell leukaemia virus (HTLV-III) disease. *Clin. Exp. Dermatol.*, **10**, 3–10
31. Eisenstat, B. A. and Wormser, G. P. (1984). Seborrhoeic dermatitis and butterfly rash in AIDS (letter). *N. Engl. J. Med.*, **311**, 189
32. Mathes, B. M. and Douglas, M. C. (1985). Seborrhoeic dermatitis in patients with acquired immunodeficiency syndrome. *J. Am. Acad. Dermatol.*, **13**, 947–951
33. Marples, R. R., Downing, D. T. and Kligman, A. M. (1972). Influence of *Pityrosporum* species in the generation of free fatty acids in human surface lipids. *J. Invest. Dermatol.*, **58**, 155–159
34. Rothman, S. (1954). *Physiology and Biochemistry of the Skin*, chap. 13. (Chicago: Chicago University Press)
35. Jarrett A. (1964). Book Review. *Br. J. Dermatol.*, **76**, 295–296
36. Kligman, A. M. (1963). In Montagna, W. (ed.) *Advances in Biology of Skin*. Vol. 4: *Sebaceous Glands*, p. 110. (Oxford: Pergamon Press)
37. Rook, A., Wilkinson, D. S. and Ebling, F. J. G. (1979). *Textbook of Dermatology*. 3rd Edn., p. 312. (Oxford: Blackwell Scientific Publications)

INDEX

www.ingramcontent.com/pod-product-compliance
Ingram Content Group UK Ltd.
Pitfield, Milton Keynes, MK11 3LW, UK
UKHW020822290926
15470UKWH00003B/1305
* 9 7 8 0 8 5 2 0 0 9 4 3 7 *